Encyclopaedia of
Nuclear Physics

Encyclopaedia of Nuclear Physics

Volume 2

Atomic Physics

Shahzad Ahmad

ANMOL PUBLICATIONS PVT. LTD.

NEW DELHI - 110 002 (INDIA)

ANMOL PUBLICATIONS PVT. LTD.
H.O.: 4374/4B, Ansari Road, Daryaganj
New Delhi - 110 002
Ph.: 23261597, 23278000
B.O.: No. 1015, Ist Main Road, BSK III Stage
III Phase, III Block
Bangalore - 560 085 (India)
Visit us at: www.anmolpublications.com

Encyclopaedia of Nuclear Physics

First Published, 2006

ISBN 81-261-2618-3 (Set)

PRINTED IN INDIA

Published by J.L. Kumar for Anmol Publications Pvt. Ltd., New Delhi - 110 002 and Printed at Mehra Offset Press, Delhi.

Contents

Preface

Science is the sermon of today. In every walk of life, since childhood till old age, science plays an important role in the life of man. Uptill some decades back, studied merely as a subject in schools and colleges, Science has assumed the status of an essential part of life today. With the expansion of literacy among the masses and growing awareness, they have woken up from the slumber of ignorance and have started taking up things more sincerely and thoughtfully. It is one of the reasons that our nation is heading towards new goals day after day.

It is a vast subject to study and talk about. For the purpose of ease, Science has been divided into more than two hundred and fifty branches. Specialisation in any branch can be obtained at different institutions and universities, scattered all over the country and abroad.

Apart from other super-specific branches, Physics, Chemistry, Zoology, Botany, Mathematics and Computronics, etc. are more general and closer to daily human life. From the point of view of the cut-throat competition in educational world, a thorough study of the above-mentioned branches, proves very handy, for the new aspirants, especially and for teachers and tutors, generally.

An urgent need was being felt for such a series of books, encompassing the commonly-studied and discussed branches of Science of equal significance and presented in an easy language, understood by all. With God's grace, our dream came true and we are happy, while introducing the dream-series for the lovers of knowledge.

The series consists of many a title, covering various branches of Science, such as Physics, Chemistry, Zoology, Botany, Geology, Agriculture Science, Computer Science, Mathematics and so on. The series is important and a must not only for students but also for teachers. Explained in simple language and aided with graphics and diagrams, everything, between the two ends—title and back cover—of the series, has been presented in a perfect way.

This set of books is devoted to **Nuclear Physics**, which comprises three volumes: (1) Nuclear Physics, (2) Atomic Physics and (3) Quantum Physics.

It is hoped that the series would serve both, the students community and the academics, equally and will prove to be a boon for those, aspiring for newer heights. Nevertheless, any suggestions for further betterment of the endeavour will be highly appreciated and solicited.

Editor

1

Atomic Physics

Molecular Physics

Orbital Magnetic Dipole Moment

Bohr Magneton : An electron revolving about a nucleus is a minute current loop and has a magnetic field. It thus behaves like a magnetic dipole.

Magnetic dipole moment,

$$|\vec{\mu}_l| = \frac{eva}{2c}\,\text{emu}$$

Orbital angular momentum, $|\vec{l}|$

$$|\vec{l}| = mva$$

Gyromagnetic ratio

$$= \frac{|\vec{\mu}_l|}{|\vec{l}|} = \frac{e}{2mc}$$

Bohr Magneton is, $\mu_B = \dfrac{eh}{4\pi m}$

Larmor Precession : When the atom is placed in an external magnetic field, the electron orbit processes about the field direction as axis. This precession is called the 'Larmor precession', and the frequency of this precession is called the 'Larmor frequency'.

Spectroscopic Terms and their Notations

- The quantised energy states of an atomic electron are described in terms of the quantum numbers n, l, s and j.
- The electron having orbital quantum number $l = 0, 1, 2, 3, 4,$ ...... are named as s, p, d, f, g.
- For example $n = 2, l = 0$ is in a 2s state

  $n = 3, l = 1$ is in a $3p$ state
- By spin orbit interaction, each energy level of a given l is splitted into two sub-levels corresponding to

  $$j = l + s = l + \frac{1}{2}$$

  $$j = l - s = l - \frac{1}{2}$$
- The number of different possible orientations of $\vec{l}$ and $\vec{s}$ and hence the number of different possible values of j is known as the 'multiplicity' of the term.

Stern-Gerlach Experiment : This experiment showed that the atoms passing through the field were oriented in space in discrete direction.

The transverse deflection suffered by the atoms is given by:

$$z = g_j m_j \frac{\frac{\partial B_z}{\partial z} \mu_B d^2}{6kT}$$

where $\frac{\partial B_z}{\partial z} \rightarrow$ rate of change of field in the z-direction.

This experiment is also an evidence for the existence of electron spin and a value $\frac{1}{2}$ is assigned to the spin quantum number $\left(s = \frac{1}{2}\right)$

Spin-Orbit Interaction : The spin-orbit interaction is an interaction between an electron's spin magnetic moment and the internal magnetic field of the atom which arises from the orbital motion of the electron through the nuclear electric field.

$$\Delta T = R_{\infty} \alpha^2 z^4 / n^3 l (l+1)$$

$$\alpha \rightarrow \text{Fine structure constant}$$

Quantum-Mechanical Relativity Correction : Besides the energy shifts due to spin orbit interaction, the relativistic effect is equally important as it produces energy shifts in the hydrogen atom comparable to those produced by spin-orbit interaction.

The relativistic term shifts is

$$\Delta T_r = -\frac{\Delta E_r}{hc}$$

$$= \frac{R_{\infty} \alpha^2 z^4}{n^3} \left(\frac{1}{l + \frac{1}{2}} - \frac{3}{4n} \right) cm^{-1}$$

Structure of Hydrogen

Lamb-Shift

- The discrepancy in the separation between the main components of the H_α line could be explained by assuming the state $2^2 S_{1/2}$ to be about 0.03 cm^{-1} higher than $2^2 P_{1/2}$ in contradiction to Dirac's theory which had shown them to have exactly the same energy.
- For hydrogen-like atoms the states of a particular *n* value having terms with the same j value but different

l values, such as $2^2 P_{1/2}$, and $2^2 S_{1/2}$, are not degenerate, but are separated. Such a separation is known as 'Lamb-Shift'.

Formulation of Pauli's Principle

- According to Pauli "no two electrons in an atom can exist in the same quantum state". This is known as Pauli's exclusion principle.

  If the particles of the system are Bose particles, the proper wave function is—

$$\psi_{Bose}(1, 2) = \frac{1}{\sqrt{2}}[\psi_a(1)\,\psi_b(2) + \psi_b(1)\,\psi_a(2)]$$

(Symmetric)

- If the particles are Fermi particles, the wave function is

$$\psi_{Fermi}(1, 2) = \frac{1}{\sqrt{2}}[\psi_a(1)\,\psi_b(2) - \psi_b(1)\,\psi_a(2)]$$

(antisymmetric)

- If the particles are in same quantum state, $a = b$,

  then $\psi_{Bose}(1, 2) \neq 0$

  but $\psi_{Fermi}(1, 2) = 0$

  Two Bose particles can exist in the same quantum state, two Fermi particles cannot.

Result of the Hartree Theory

1. The electron for the average value of r (radius of shell) holding for a multi-electron atom would be

$$\bar{r} = \frac{n^2 a_0}{z_n}\left[1+\frac{1}{2}\left\{1-\frac{l(l+1)}{n^2}\right\}\right] \simeq \frac{n^2 a_0}{z_n}$$

In multi-electron atoms the radii of the inner shells of small n are very small, e.g., the radius of the $n = 1$ shell of argon ($z_n \simeq z - 2 \simeq 16$) is very smaller than that of the $n = 1$ shell of hydrogen by approximately a factor of $\frac{1}{16}$.

2. For the outermost populated shell of an atom, Hartree calculations show that

$$z_n \simeq n$$

and so the corresponding radius is

$$\bar{r} \simeq \frac{n^2 a_0}{z_n} \simeq \frac{n^2 a_0}{n} \simeq n a_0$$

Thus the radius of the outermost shell of a multi-electron atom is roughly n times, the radius of the smallest orbit of hydrogen.

3. Like radius, we can approximately express the total energy of an electron in a multi-electron atom by the energy equation of a one-electron atom, where z is replaced by z_n, *i.e.*

$$E \simeq -\frac{2\pi^2 \mu z_n^{\,2} e^4}{n^2 h^2}$$

4. For the outermost populated shell of an atom, we have $z_n \simeq z$ and corresponding energy is

$$E \simeq -\frac{2\pi^2 \mu e^4}{h^2}$$

Thus, the total energy of an electron in the outermost shell of any atom is comparable to that of an electron in the ground state of hydrogen atom.

5. The energy equation for a multi-electron atom contains $z_n^{\,2}$ in the numerator and n^2 in the denominator. Meaning: for a given atom the total energy E becomes less negative with increasing n.

6. In a one-electron atom the total energy E depends only on the quantum number n, while in a multi-electron atom it depends on l as well as on n.

Spectroscopic Terms : L-S and j-j Couplings

Terminology

State : The 'state' of an atom is the condition of motion of all the electrons specified by four quantum numbers.

Degenerate : If several states have the same energy.

Ground State : The state with lowest energy is the ground state.

Energy Level : A collection of states having the same energy in the absence of external magnetic or electric field constitutes an "energy level". The level with the lowest energy is the ground level.

Sub Level : An external field splits an energy level into several "sublevels" each characterised by one or more magnetic quantum number.

Term : A collection of levels characterised by an orbital angular momentum and multiplicity comprises a 'Spectroscopic term'.

e.g. : a^3D term means the weighted average energy of the $^3D_3$, $^3D_2$, and $^3D_1$ levels.

Configuration : The specification of the quantum numbers n and l for the orbitals of all the electrons of an atom is called the 'electron configuration' of the atom. *e.g.* the electron configuration of the ^{6}C atom : $1s^2\ 2s^2\ 2p^2$.

Equivalent Orbitals : Orbitals with the same n and l are 'equivalent'. The electrons in equivalent orbitals are called 'equivalent electrons'.

Statistical Weight : The number of distinct states in a specified collection is the 'statistical weight'. The statistical weight of a level is $2J + 1$; for term it is $(2S + 1)(2L + 1)$; for a single electron it is $2n^2$.

Spectral lines are categorised according to the following nomenclature.

Component : A transition between two sublevels is called a 'Component'.

Line : A transition between two levels is a 'line'. A line is thus a blend of components.

Multiplet : A collection of transitions between two term is called a multiplet. A multiplet thus consists of a number of lines.

Resonance Line : Among the lines arising from transition between the ground level and higher levels, the line of lowest frequency is called the "resonance line".

Selection Rules for Multi-electron Atoms in L-S Coupling: In the electric-dipole transitions in multi-electron atoms, the selection rules are—

1. $\Delta l = \pm 1$ (for one electron)

   The parity of the configuration must change in an electric dipole transition (odd $\leftrightarrow$ even).

   For two electron transition we would have

   $$\Delta l_1 = \pm 1 \; ; \; \Delta l_2 = 0, \pm 2$$

2. There is no restriction on the total quantum number n of either electron.

3. For the atom as a whole, the quantum numbers L, S and J must change as follows:

   $\Delta L = 0, \pm 1$ (In one-electron atoms $\Delta L = 0$ is not allowed)

   $$\Delta S = 0$$

   $$\Delta J = 0, \pm 1 \text{ but } J = 0 \not\leftrightarrow J = 0.$$

j - j Coupling

(a) As a result of the stronger spin-orbit interaction, the orbital and spin angular momentum vectors of each individual electron are strongly coupled together to form a resultant angular momentum vector $\vec{j}$ of magnitude $\sqrt{j(j+1)}\dfrac{h}{2\pi'}$ where $j = l-\dfrac{1}{2}$ and $l+\dfrac{1}{2}$, *i.e.*, j takes half integral values only.

(b) As a result of the residual electrostatic interaction and spin-spin corelation, the resultant angular momentum vector $\vec{j}$ of the individual electrons are less strongly coupled with one another to form the total angular momentum vector $\vec{j}$ of the atom, of magnitude $\sqrt{j(j+1)}\dfrac{h}{2\pi'}$ where the quantum number j has the values.

$$J = |\vec{j_1} + \vec{j_2} + \ldots|_{min}, |\vec{j_1} + \vec{j_2} + \ldots|_{min} + 1, \ldots \ (j1 + j2 + \ldots)$$

Selection Rules in j - j Coupling

1. If two electron jump then

$$\Delta l_1 = \pm 1, \Delta l_2 = 0, \pm 2$$

2. $\Delta j = 0, \pm 1$ for the jumping electron and $\Delta j = 0$ for all the other electrons.

3. For the atom as a whole

$$\Delta J = 0, \pm 1 \text{ but } J = 0 \not\leftrightarrow J = 0$$

The selection rules $\Delta S = 0$ and $\Delta L = 0, \pm 1$ no longer hold, since L and S are no longer good quantum numbers.

The Zeeman Effect

- When an atom is placed in an external magnetic field, the spectral lines it emits are split into several polarised components. This effect of magnetic field on the atomic spectral lines is called Zeeman effect.

- A singlet spectral line viewed normal to the field is split into three plane-polarised components ; a central unshift line with the electric vector vibrating parallel to the field (called π component) and two other lines equally displaced one on either side with electric vector perpendicular to the field (called σ components). This is called a "Normal triplet" and effect is 'Normal Zeeman effect'.

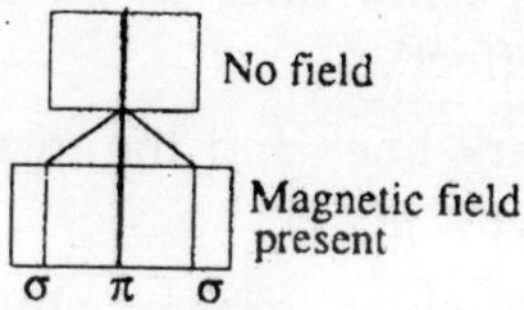

Normal Triplet

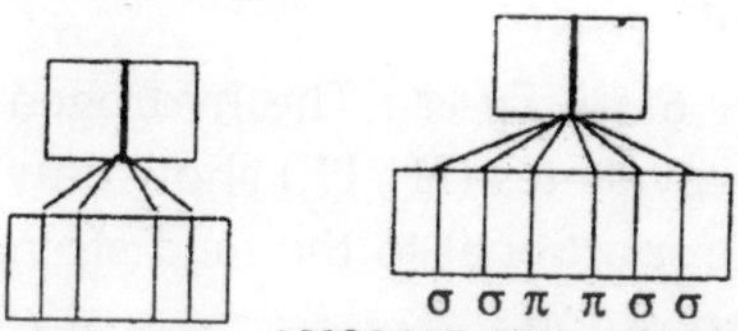

ANOMALOUS PATTERNS

- The fine-structure components of a multiplet spectral line, however, show a complex Zeeman pattern, e.g., D_1 and D_2 components of sodium yellow doublet give four and six lines respectively in the Zeeman pattern. This is 'anomalous' Zeeman effect.

  The Landed-factor is given by

$$g = 1 + \frac{J(J+1) + S(S+1) - L(L+1)}{2J(J+1)}$$

Paschen-Back Effect : The strength of the external field is increased, the separations between zeeman components increase until they become greater than the separations between multiplet fine-structure components. The anomalous Zeeman pattern then changes over to like a normal Zeeman pattern. This phenomenon is known as 'Paschen-Back effect'.

In a strong field, the selection rules for transitions are

$$\Delta M_L = 0$$

(Components polarised parallel to the field)

$$\Delta M_L = \pm 1$$

(Components polarised perpendicular to the field)

$$\Delta M_S = +1$$

Stark Effect

- The splitting of spectral lines in an electric field is known as 'Stark effect'.
- The number of stark lines and the total width of the pattern increase with *n*.
- The π-components show greater shifts than the σ component.
- *First Order Stark Effect* : The hydrogen lines with the lower energy states (H_{α}, H_{β}) show only a symmetrical splitting proportional to the field strength about their field free position.
- *Second Order Stark Effect :* For the lines involving the higher states (H_{γ}, H_{δ}), the Stark components show unidirectional displacements proportional to the square of the field strength.
- The Stark-Shift for the ground state ($n = 1$) is zero.
- For a non-degenerate state, such as the ground state ($n = 1, l = 0, n_1 = 0$) of hydrogen atom which has even parity, there is no first order Stark effect.

  Only degenerate terms can have a first order Stark effect.

Nuclear Spin and Hyperfine Splitting

- Bismuth exits as a single isotope, but shows six hyperfine components in its line at wevelength 4722Å.

The nuclear angular momentum is

$$|\vec{I}| = \sqrt{I(I+1)}\frac{h}{2\pi}$$

I → Nuclear spin quantum number.

Nuclear magnetic moment.

$$\mu_1 = g_I \sqrt{I(I+1)}\mu_N$$

$$\left[\mu_N = \frac{eh}{4\pi m_p c}\right]$$

- We have only to count the number of hyperfine components for any Zeeman line in a strong magnetic field, and this number is 2I + 1. Hence I is directly determined. Back and Goudsmit found that in a strong magnetic field each of the Zeeman components of Bi consists of 10 hyperfine components due to nuclear spin. Hence they concluded that for B_i, $I = \frac{9}{2}$.

X-Rays

- X-rays are produced when high speed electrons strike a metal of high atomic number.

X-rays spectrum consists of two spectra :

1. *Continuous Spectrum* : It is due to continuous deacceleration of the electron as it passes through the field of the nucleus. Its short-wevelength limit is given by—

$$\lambda_{min} = 12{,}400/V \text{Å}$$

2. *Line or Characteristic Spectrum* : It is produced when electrons of the target material are dislodged from their orbits by the striking electrons. The spectrum consists K-, L- and M-series etc. depending on when electron jump ends.

Moseley's Law : Moseley's law states that the frequency of a spectral line in the characteristic x-ray spectrum varies directly as the square of the atomic number of the element emitting it.

- Absorption of X-rays follows the exponential law

$$I = I_0 e^{-\mu x}$$

  where μ is the absorption coefficient.

- The linear absorption coefficient μ may be defined as the fractional decrease in the intensity of the X-rays per unit thickness of the absorber material.
- Due to Compton scattering, change in wavelength of the scattered X-rays is given by—

$$\Delta\lambda = 0.024\ (1 - \cos\theta)\ \text{Å}$$

Molecular Spectra

The Born-Oppenheimer Approximation

- We cannot classify molecular energy state according to the electronic angular momentum L which is not conserved in molecules.
- The energy of each of the vibrational and rotational motions is quantised with the result that there are many more energy levels in a molecule than in an atom.
- We can solve the complete molecular wave equation in the following way.

  1. The electronic wave equation $H_e\psi_e = E_e'\psi_e$ for a given electronic state of the molecule is solved for a range of values of nuclear coordinates. This gives ψ_e and E_e as continuous functions of nuclear co-ordinates.

2. Having obtained E'_e the nuclear wave equation $(H_n + E_e') \psi_n = E\psi_n$ is solved. The eigen values E are the characteristic energies for the whole molecule in the given electronic state.

Types of Molecular Spectra

Electronic Spectra : These spectra are observed, both in emisson and absorption in the visible and ultraviolet regions. The electronic spectra are observed for the heteronuclear as well as for the homonuclear diatomic molecules.

Vibrational-rotational Spectra : These spectra are observed in absorption, in the near infrared region ($1\mu - 10^2\ \mu$). They are observed only for the heteronuclear molecules. The homonuclear molecules like H_2, N_2, O_2, do not produce vibrational-rotational band.

Pure Rotational Spectra : These spectra are observed in absorption in the far infrared ($10^2\mu - 10^3\mu$) and observed only for the heteronuclear diatomic molecules.

Pure Rotational Spectra : Only molecules that have permanent electric dipole moments can absorb or emit electro-magnetic radiation in such transition.

Hamonuclear diatomic molecules such as H_2, O_2, N_2 ; symmetric linear molecules such as CO_2 and spherical top polyatomic molecules such as CH_4 do not exhibit rotational spectra.

- The moment of inertia of the molecule as a rigid rotator

$$I = \mu r^2$$

$\mu \rightarrow$ Reduced mass ;

$r \rightarrow$ distance from the axis of rotation.

- Eigen values, E are

$$E = \frac{h^2}{8\pi^2 I} J(J + 1)$$

J → Rotational quantum number;

$$J = 0, 1, 2, 3,...$$

- In terms of wave number, the energy equation can be written as

$$\frac{E}{hc} = F(J)$$

$$= BJ(J + 1)\text{cm}^{-1} \text{ with}$$

$$B = \frac{h}{8\pi^2 Ic}$$

- Wave number

$$v = 2B(J + 1)$$

with $J = 0, 1, 2,...$

$$v = 2B, 4B, 6B,...$$

Thus, the absorption spectrum of a rigid rotator is expected to consists of a series of equidistant lines with constant separation 2B.

- The population in the Jth level relative to the lowest level ;

$$\frac{N_J}{N_0} = (2J + 1)\, e^{-(E_J - E_0)/kT}$$

$$= (2J + 1)\, e^{-BJ(J+1)hc/kT}$$

Vibrational Rotational Spectra : In the series of lines which are not equidistant, a line is missing in the centre of the band. The missing line is known as the 'null line' or 'zero gap'.

The Molecule as a Harmonic Oscillator : The allowed energies for the harmonic oscillator $E = h\upsilon_{osc}\left(v + \frac{1}{2}\right)$

where $v = 0, 1, 2,......$

- The vibrational terms

$$G(v) = \frac{E}{hc} = \frac{v_{osc}}{c}\left(v + \frac{1}{2}\right) = \omega\left(v + \frac{1}{2}\right)$$

where $\omega = \frac{v_{osc}}{c}$ → Vibrational constant.

- Vibrational transitions can only occur when the molecule has a permanent dipole moment which changes with distance.
- The spacing between the vibrational levels are considerably larger than the spacings between the rotational levels of a molecule, intact larger than kT at room temperature.
- Thus, the main vibrational transition in absorption is

$$v = 1 \leftarrow v = 0$$

- The vibrational spectrum is expected to consists of a single band at ω cm^{-1}.
- In the vibrational spectrum, the nuclei carrying out harmonic vibrations along the internuclear axis.

Molecule as Anharmonic Oscillator

- The actual infrared spectrum is found to consists at an intense (fundamental) band at ω plus a number of weak bands (overtones) at slightly lesser and lesser than 2ω, 3ω, ... The observation of overtones indicates that the selection rule $\Delta v = \pm 1$ is not strictly obeyed and transitions corresponding to $\Delta v > 1$ do take place.
- The observation that the overtones appear not exactly at 2ω, 3ω, ... but at lesser and lesser values indicates that the vibrational energy levels are not exactly equally-spaced but converges slowly.

- The corresponding term values are given by

$$G(v) = \omega_e\left(v+\frac{1}{2}\right)-\omega_e x_e\left(v+\frac{1}{2}\right)^2+...$$

$\omega_e x_e \rightarrow$ the anharmonicity constant smaller than ω_e and always positive.

The vibrational frequency of anharmonic oscillator is

$$\therefore \quad v_{osc}(v) = c\,[(\omega_e - \omega_e x_e) - 2\omega_e x_e\, v]$$

The frequency

$$V_{osc}\left(-\frac{1}{2}\right) = c\omega_e\ [v = -\frac{1}{2} \text{ is hypothetical state}]$$

$\omega_e \rightarrow$ The vibrational frequency that anharmonic oscillator would have classically for an infinitesimal amplitude *i.e.*, in imaginary state $v = -\frac{1}{2}$ at the very bottom of the potential curve.

$$v_{osc} = c\omega_e = \frac{1}{2\pi}\sqrt{\frac{k_e}{\mu}}$$

$$\text{or} \quad k_e = 4\pi^2\mu c^2\ \omega_e^2$$

Isotop Effect on Vibrational Level : The vibrational levels of the heavier isotope are lower than the corresponding levels of the lighter isotope.

$$G'(v) < G(v)$$

Molecule as Vibrating Rotator

Fine Structure of Infrared Bands : The near infrared spectra of molecules consists of 'band' (not lines) each band being composed of close lines arranged in a particular manner. This observed fine structure suggests that in a vibrational transition the molecule also changes its rotational energy state, *i.e.*, it must be treated as a vibrating-rotator.

- B, determines the spacing between the rotational levels, is different for different vibrational states.
- For a given vibrational transition, the rotational transitions $\Delta J = +1$ given one set of lines called the 'R-branch' while the rotational transitions $\Delta J = -1$ give the other set of lines called the 'P-branch'.
- $V_0 = G(v') - G(v'')$, is the wave number of the pure vibrational transition (for which $J' = J'' = 0$) which is unallowed ($\Delta J \neq 0$) and corresponds to the missing line in the band. V_0 is known as the wave number of the 'band-origin'.
- If we neglect vibration-rotation interaction, then

$$V_R = V_0 + 2B + 2BJ''$$

[Wave number of R branch]

$$V_P = V_0 - 2BJ''$$

[Wave number of P branch]

- The only case of a diatomic molecule for which a Q-branch (*i.e.* a central 'line') has been observed in the rotation-vibration infrared bands is that of nitric oxide NO.

Electronic Spectra : Franck-Condon Principle

- Each electronic states of the molecule has a different potential energy curve.
- A single electronic transition in a molecule gives rise to a band system.

Electronic Band Spectra in Absorption : The study of absorption spectra, as compared to emission spectra, gives more information, with less elaborate computation, regarding the excited electronic states of molecules.

- The absorption spectrum of I; molecule, show progressions corresponding to $v'' = 1$ and even higher, in addition to the progression $v'' = 0$, although with smaller intensity.
- The P and R branch can be represented by a single parabolic equation.

$$v = v_0 + (B_v' + B_v'')m + (B_v' - B_v'')m^2$$

where

$$m = J'' + 1 = 1, 2, 3, \ldots\ldots$$

for lines R(0), R(1), R(2),...

$$m = -J'' = -1, -2, -3, \ldots\ldots$$

for lines P(1), P(2), P(3),...

$$m = 0 \text{ for the band origin.}$$

Franck Condon Principle : A vibrational transition will have a large Franck-Condon factor only if one of the two turning points of a level of one electronic state lies at approximately the same internuclear distance as one of the two turning point of a level of the other electronic state, except in case of $v = 0$ level for which the middle-point of the level rather than the turning point must be substituted.

- The potential curves become increasingly asymmetrical with increasing r. This means that the nuclei spend more time in the extended phase ($r > r_e$) than in the contracted phase ($r < r_e$). Hence according to Franck-Condon principle, transition will be more probable from (or to) the one side than the other.

∴ for a system degraded to the red ($r' > r''$) the right limb of the Condon parabola will be the stronger.

2

Atomic Shells

Continuous and Characteristic X-rays : When a beam of X-ray emitted by the target of an X-ray tube is analysed by a Brag's crystal spectrometer, two distinct types of spectra are obtained :

(*i*) a continuous spectrum, and

(*ii*) a line spectrum superposed on the continuous spectrum.

This shows that the X-ray beam from the target is made up of a number of continuous wavelengths of varying intensity; and that the intensity rises sharply at certain definite wavelengths.

Fig. Illustrates roughly the variation of intensity of the beam with wavelength λ when the X-ray tube is operated at a sufficiently high voltage. The continuous spectrum has a definite snort wavelength limit $\lambda_{min,}$ *below* which there is no radiation. This limit is independent of the material of the target and depends on the voltage applied to the X-ray tube. The radiation having the continuous wavelengths forms the `continuous X-rays'.

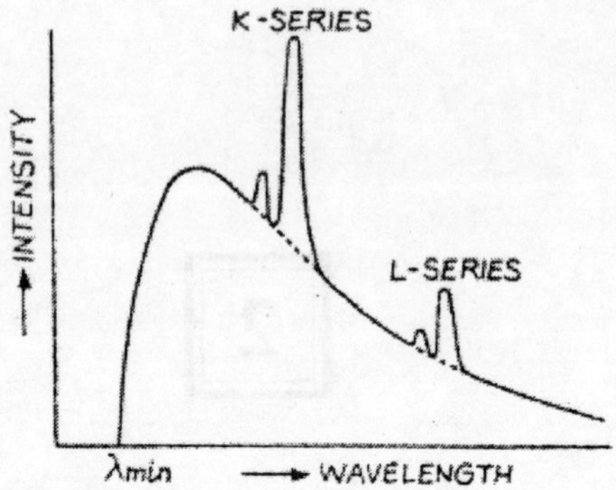

The continuous curve has pronounced intensity peaks at certain wavelengths, indicating the enhanced production of X-rays. The positions of these peaks, however, depend only on the material of the target. They occur in groups known as the *K, L, M*, ..., series. These series of lines are known as the X-ray spectra of the material of the target. The radiation having definite wavelengths which are characteristic of the material of the target forms the 'characteristic X-rays'.

Origin of Continuous X-rays: The emission of continuous spectrum can be explained by the classical theory. When the electrons emitted from the cathode of the X-ray tube strike the target, they are deflected by the strong fields of force of the nuclei of the target atom. Such electrons, according to classical theory, emit pulses of radiation of all possible wavelengths which form the continuous spectrum. The theory, however, fails to explain the occurrence of the short wavelength limit, which requires quantum theory for its explanation.

Short Wavelength Limit and Accelerating Voltage, Duane-Hunt Law: According to the quantum theory, the radiation is emitted (or absorbed) in discrete packets of energy or quanta, called 'photons', the energy of each photon being $h\nu$, where ν is the frequency of the radiation and h is 'Planck's constant'. Now, when an electron of charge e emitted from the cathode is accelerated through the voltage V applied to the X-ray tube, it acquires an amount of kinetic energy eV. When this electron collides an atom of the target, it may lose most or all of its energy, which is re-emitted as an X-ray photon.

For the most favourable collision in which the electron loses the whole of its energy in a single collision with the target atom, an X-ray photon of maximum energy, $h\upsilon_{max}$, is emitted. Thus for an accelerating voltage V, the maximum X-ray photon energy is given by

$$eV = h\upsilon_{max}.$$

Consequently, the maximum frequency of X-rays produced by electrons of energy eV is

$$V_{max} = \frac{eV}{h}.$$

A maximum frequency υ_{max} corresponds to a minimum wavelength λ_{min}, such that

$$\lambda_{min} = \frac{c}{\upsilon_{max}},$$

where c is the speed of light. Therefore, the minimum wavelength of the X-rays for a given voltage V across an X-ray tube is

$$\lambda_{min} = \frac{hc}{eV}.$$

This is Duane-Hunt relation. Putting the valves :

$$h = 6.63 \times 10^{-34} \text{ J-s},$$

$$c = 3.0 \times 10^{3} \text{ m-s}^{-1}$$

and $e = 1.6 \times 10^{-19}$ C, we get

$$\lambda_{min} = \frac{1.24 \times 10^{-6}}{V} \, m = \frac{12400}{V} \text{ Å}.$$

Thus λ_{min} is inversely proportional to V.

Duane and Hunt experimentally studied the variation of

intensity with wavelength in the continuous spectrum at different accelerating voltages applied to the X-ray tube (Fig.). For each voltage, the intensity curve starts at a particular minimum wavelength, rises rapidly to a maximum and then drops gradually but indefinitely towards the longer wavelengths. The wavelength at which the intensity is maximum depends on accelerating voltage, being shorter for higher voltages. Similarly, the short wavelength limit decreases as the accelerating voltage increases. Infact, Duane and Hunt found that the short wavelength limit varies inversely as the accelerating voltage. Thus, the short wavelength limit of continuous X-rays provides a confirmation of quantum theory.

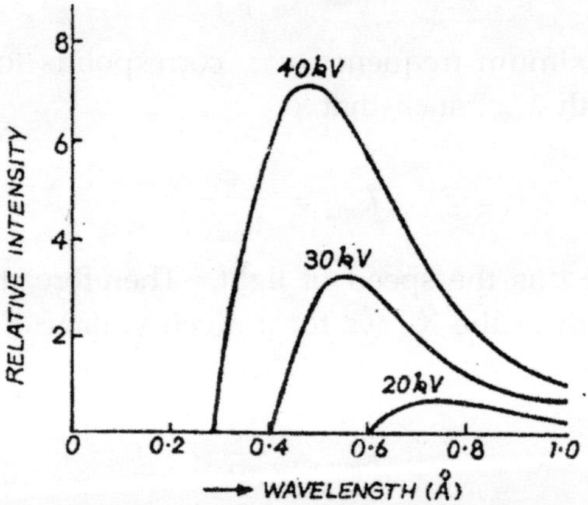

The electron rarely loses whole of its energy in a single collision. More usually it undergoes a sequence of collisions with atoms before coming to rest, thus emitting photons of smaller and smaller energies, *i.e.* of larger and larger wavelengths. This accounts for the production of a continuous range of X-ray wavelengths above a definite limit.

In fact most of the electrons falling upon the target lose their kinetic energy gradually in numerous collisions, thus merely increasing the average kinetic energy of translation of the atoms of the target. The result is that the temperature of the target is increased. Thus most of the kinetic energy of the electron-beam striking the target goes simply into heat. It is only 1% or even less of the electron-energy which is converted into X-rays.

Characteristic X-ray Line Spectrum

When the target of an X-ray tube is struck by energetic electrons, it emits two kinds of X-ray radiation. One of them has a continuous spectrum whose intensity distribution in wavelength depends entirely on the tube voltage. The other consists of distinct spectral lines whose frequencies are characteristic of the material of the target. Hence this line spectrum is called the X-ray spectrum of the material of the target. The radiation forming the line spectrum is called as 'characteristic X-rays'. The frequencies of the characteristic X-rays are, in general, about 1000 times higher than the optical frequencies.

The characteristic *X*-ray spectra are distinguished by their simplicity. Most elements show in their *X*-ray spectra two series of lines known as *K*-series and L-series. Each series contains a small number of lines designated as $K\alpha$, $K\beta$.......$L\alpha$ $L\beta$.......The wavelengths of *K*-series are generally less than 1 Å, while those of *L*-series are roughly ten times greater. Heavier elements ($Z > 66$) show further series known as M-, *N*-, *O*-, series.

Kossel's Explanation

Kossel explained the origin of the characteristic X-ray line spectrum on the basis of the shell structure of the atom. An atom is built up of a central positively-charged nucleus with discrete aggregates of electrons, known as *K*-shell, *L*-shell, *M*-shell, *N*-shell, etc. The maximum number of electrons which the *K, L, M, N,*......shells can hold are 2, 8, 18, 32,....... The electrons in the *K*-shell are attracted by the positive nucleus with the greatest force, and to eject them from the atom maximum energy is required. Lesser energy is required to eject an electron from the L-shell, still lesser from *M*-shell, and so on.

Now, in an X-ray tube an electron emitted from the cathode strikes the target with a tremendous velocity so that it penetrates well inside the atoms of the target. If it ejects an electron from the *K*-shell of the atom, a vacancy is created in the *K*-shell. Immediately an electron from one of the outer shells, say *L*-shell, jumps to the *K*-shell, emitting an *X*-ray photon of energy equal to the energy difference between the two shells. Similarly, if an electron from the *M*-shell jumps to the *K*-shell, X-ray photon of higher energy is emitted. The *X*-ray photons emitted due to the jump of electron from the *L, M, N,...*shells to the *K*-shell give *K*α, *K*β, *K*γ... lines of the *K*-series of the spectrum.

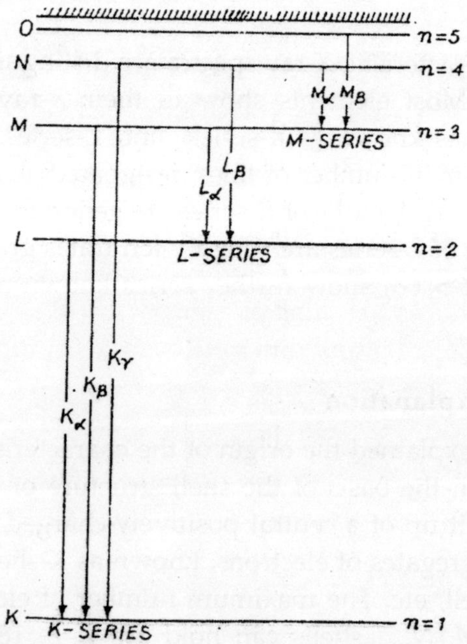

If the electron striking the target ejects an electron from the *L*-shell of the target atom, an electron from the *M, N*...shell jumps to the *L*-shell, so that X-ray photons of lesser energy are emitted. These photons form the *L*-series of the spectrum. Clearly, the frequencies of the L-series are smaller than those of the *K*-series. In a similar way, the formation of *M*-series, *N. series* etc. may be explained. The energy level diagram is shown in Fig. The *K, L,* M,...shells correspond to quantum numbers n = 1, 2, 3.....

Whenever a vacancy is created in an inner shell, say *K*-shell, the electron from the next higher shell, the *L*-shell, is most likely to jump to the *K*-shell emitting a line of the *K*-series. The vacancy left by the jumping electron in the *L*-shell is filled by the jump of another electron from a still higher shell, so that a line of *L*-series is emitted. This process continues until the vacancy reaches the outermost shell where it is filled by some free electron. Thus the emission of *K*-series is always accompanied by the emission of *L*-, *M*–, *N*- ... series.

Moseley's Law

Moseley, in 1913, made an extensive study of the characteristic X-ray spectra of a number of heavy elements, and observed a simple relationship between them. He found that the spectra of different elements are very similar, and with increasing atomic number Z, the spectral lines merely shift towards shorter wavelengths or higher frequencies (Fig.). He plotted a graph of the K-series between the square-root of frequency $\left(\sqrt{v}\right)$ and atomic number (Z) and found it very approximately to be a straight line (Fig.). He therefore concluded that the square-root of the frequency of a K-line is closely proportional to the atomic number of the element. This is called Moseley's law and may be expressed as

$$\sqrt{v} = k\ (Z - A),$$

where Z is the atomic number of the element, and k and b are constants for a given transition of the *K*-series. The law is applicable to other series also but with different values of k and b.

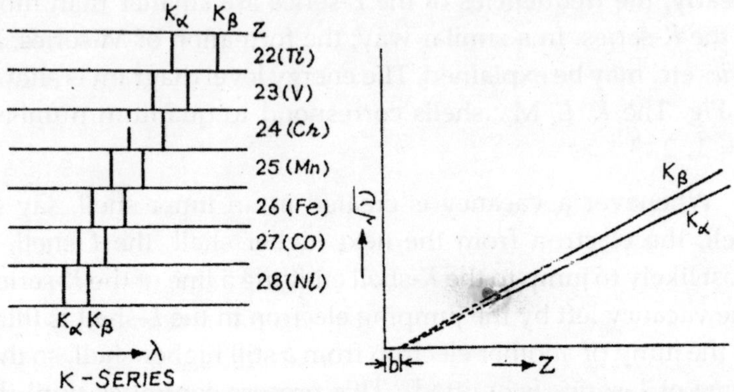

If, however, a graph be plotted between $\sqrt{v}$ and the atomic weight, there is an appreciable departure from straight line. Moseley therefore concluded that it is the atomic number, and not the atomic weight, which is more fundamental with regard to the emission of characteristic X-rays.

Explanation of Moseley's Law from Bohr Theory: At about the same time when Moseley gave his law regarding characteristic X-ray spectra, Bohr presented his theory of the spectrum of the hydrogen atom. It then became evident that Bohr's theory could explain characteristic X-ray spectra, if the atomic number Z was identified as the number of (positive) charges on the nucleus. According to Bohr theory, the wavelength λ of the hydrogen spectral line emitted when an electron jumps from an initial energy level n_i to a final energy level nf is given by

$$\frac{1}{\lambda} = RZ^2\left(\frac{1}{n_{f^2}} - \frac{1}{n_{i^2}}\right), \quad \text{... } (i)$$

where R is Rydberg's constant. Now, in X-ray emission if one of the two K-electrons of an atom is removed, an L-electron would be attracted by the positive charge of the nucleus in very much the same way as the single electron of a hydrogen atom is attracted by its nucleus. Since the remaining *K*-electron is much nearer the nucleus than the *L*-electron; it produces a screening effect reducing the attraction of the nucleus for the *L*-electron. Thus for the *L*-electron the effective nuclear charge is reduced to (Z – 1) *i.e.* replacing Z by (Z – 1) in the Bohr formula (i), and remembering that when an *L*-electron jumps into a *K*-shell, the quantum number n changes from 2 to 1 (*i.e.* $n_t = 2$ and $n_f = 1$) we find that the wavelength of K_α line of an element of atomic number Z is given by

$$\frac{1}{\lambda} = R(Z-1)^2\left(\frac{1}{1^2}-\frac{1}{2^2}\right).$$

The line L_α arises when an *M*-electron jumps into the *L*-shell having a vacancy of one electron (*i.e.* $n_i=3$ and $n_f= 2$). Now, the nine electrons remaining in the *K* and *L* shells shield the nucleus during the jump of the electron from the *M* to the *L* shell. However, this shielding is not perfect and the effective Z for the emission of L_α line is reduced only to (Z—7.4). Thus the wavelength of L_α-line is given by

$$\frac{1}{\lambda} = R\,(Z-7.4)^2\left(\frac{1}{2^2}-\frac{1}{3^2}\right).$$

In general, we can write

$$\frac{1}{\lambda} = R\,(Z-b)^2\left(\frac{1}{n_{f^2}}-\frac{1}{n_{i^2}}\right),$$

where b is screening constant.

The formula for the frequency v (— c/λ) can be written as

$$v = Rc\,(Z-b)^2\left(\frac{1}{n_{f^2}}-\frac{1}{n_{i^2}}\right),$$

$$= k^2\,(z-b)^2$$

$$\sqrt{v} = \mathrm{k}\,(Z-b),$$

which is Moseley's law.

Importance of Moseley's Law with Regard to the Periodic Table: Mendeleev had arranged the elements in the periodic table in the order of increasing atomic weight. This table had a few anomalies. For example, the properties of Argon, Cobalt and Tellurium were not in agreement with the properties of other elements of their groups.

Moseley's work on X-ray spectra indicated that an element is distinguished by the number of positive charges on the nucleus of its atom *i.e.* by the atomic number Z. Rutherford in the same year showed that the atomic number was the fundamental constant which decided the chemical properties of the atom. Moseley then pointed out that the elements in the periodic table must be arranged in the order of increasing atomic number (instead of atomic weight). From this point of view he changed the positions of certain elements in the Mendeleev's periodic table. For instance, he found that the atomic number of Cobalt ($_{27}Co^{58.9}$) is less than that of nickel ($_{28}Ni^{58.7}$) even though its atomic weight is greater. Therefore he arranged them in the order Co, Ni instead of Ni, Co. Similarly, he placed argon ($_{18}A^{40}$) before potassium ($_{19}K.^{39}$), and tellurium ($_{52}Te^{127.6}$) before iodine ($_{53}I^{127}$). When these changes were made, the anomalies of the Mendeleev's table disappeared.

While arranging the elements in the order of increasing atomic number, Moseley had to left certain gaps such as at Z = 43, and 72. He pointed out that the corresponding elements could exist but had not then been discovered. These elements were subsequently discovered : they are technetium and hafnium respectively.

X–ray Spectra and Optical Spectra

The X-ray spectra emitted by different elements are very like one another and much simpler than the corresponding optical (visible) spectra. The main differences between the two spectra are as follows :

(*i*) The frequencies in X-ray spectra are about 1000 times higher than those in optical spectra. This indicates that X-ray spectra arise from electron transitions among the inner complete shells of very firmly bound electrons. This is also in agreement with the fact that the X-ray spectra of a free element is same as that of its chemical compounds. The optical spectra, on the other hand, arise from transitions of the outermost electrons of the atom.

(*ii*) The X-ray spectra vary smoothly from element to element. This is because the X-ray spectra depend on the binding energies of the electrons in the inner shells. These energies simply increase uniformly from element to element owing to the increasing nuclear charge. The optical spectra, on the other hand, show abrupt periodic changes from element to element. This is because the electronic structure at the surface of the atom changes periodically from element to element.

(*iii*) The X-ray spectra arise as a result of the complete removal of an electron from an inner shell of the atom. In case of optical spectra the electron from an outer shell is simply raised to a higher level and is not completely removed.

(*iv*) The X-ray emission spectra are entirely different from the X-ray absorption spectra. The former consists of discrete lines while the latter consists of continuous regions bounded by sharp edges. On the other hand, the optical emission spectra are identical to the optical absorption spectra.

It is found that the X-ray spectra of different elements are qualitatively very similar. With increasing atomic number, *i.e.* from element to element the characteristic X-ray lines merely shift towards shorter wavelengths or higher frequencies. There are no abrupt changes from one element to the next. The reason is that X-ray spectra arise from electronic transitions among the inner closed shells of the atom. The electronic arrangement in the innermost shells is regular and essentially the same in all elements (except in very light elements). The nuclear charge, however, increases steadily from element to element and this results in the steady increase in the frequency of the corresponding X-ray lines (Moseley's law).

The optical spectra arise from electronic transitions among the outer shells of the atom. The electronic arrangement in these shells changes periodically and rather abruptly from element to element. Hence the optical spectra also change periodically and abruptly.

X-ray Absorption Spectrum

Absorption Edges: When a beam of continuous X-rays is passed through the film of a substance and the transmitted beam is sent into an X-ray spectrograph, an X-ray absorption spectrum of that substance is obtained. The nature of this spectrum is quite different from the X-ray emission spectrum of the same substance. None of the emission lines K_α, K_β ...appears in the absorption spectrum. Instead the absorption spectrum shows a continuous region bounded by a sharp edge in the position of the limit of the *K* series. When successive elements in the periodic table are used as absorbers, the edge shifts toward higher frequency (Fig.). Thus these edges are characteristic of the absorber element and are called 'absorption edges'.

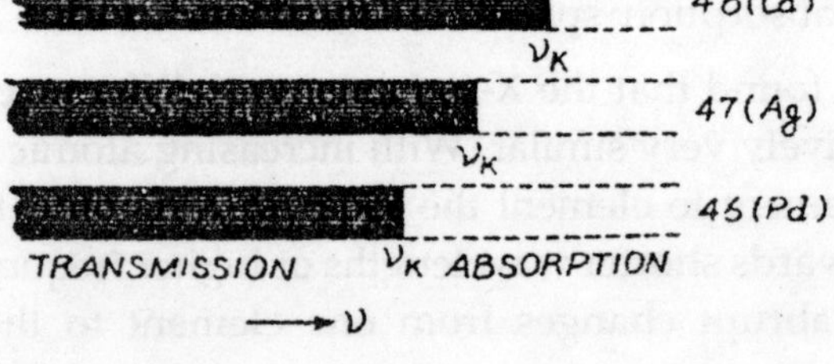

***Explanation*:** These are two features of the X-ray absorption spectrum which are to be explained:

(i) *Absence of Absorption Lines:* The K_α-line (for example) in emission, occurs when an electron drops from the *L*-shell to the *K*-shell. In absorption this line would occur only if an electron from the *K*-shell, after absorbing an X-ray photon of the required energy from the continuous beam, could *go* to the *L*-shell. But this does not happen because the *L*-shell has already its full quota of 8 electrons and so it cannot receive any further electron Pauli's exclusion principle). Hence the K_α-line cannot be observed as absorption line. This is true for all the X-ray emission lines that are commonly observed.

(ii) *Appearance of Absorption Edges:* An X-ray photon can, however, be absorbed if it has enough energy to remove an inner electron to infinity where the shells no longer exist. If W_λ- is the energy required to remove a *K*-electron, then a photon of frequency v can eject a *K*-electron provided $v \geq vk$ where

$$hv_K = W_K$$

Photons of frequency smaller than v_K would be transmitted by the absorbing element.

The black part in Fig. corresponds to the transmission of X-ray photons of all frequencies less then v_K. At the frequency v_K, absorption in the *K*-shell suddenly starts and thereafter this absorption continues for all frequencies greater than v_K. Hence the absorption spectrum is abruptly discontinued at v_K (no blackening of photographic plate beyond v_K).

As Z increases, W_K increases, and so V_k also increases. Hence for successive elements in the periodic table the absorption edge shifts towards higher frequency. The frequency V_k (or the wavelength λ_K) at which absorption in the *K*-shell begins is called the '*K* absorption limit' of the particular element.

Thus *the K-absorpt on edge observed in X-rays, correspond to the photoelectric ejection of an electron from the K-shell.* Absorption beyond the edge indicates that the *K*-electron passes to infinity with the energy E_K given by Einstein's photoelectric equation ($E_K = h\nu — W_K = h\nu - h\nu_K$). Since each element gives only one *K*- absorption edge, we conclude that the *K*-shell has only a single energy level.

The Multiplicity

When the X-ray absorption spectrum is taken in the *L*-region, again a continuum is observed but now with three absorption edges. This means that the *L*-shell has three energy levels from which electrons can be ejected to infinity. Similarly, .M-shell has five energy levels, and so on. Thus X-rays absorption spectrum reveals the multiplicity of atomic shells.

PROBLEMS

1. *What are the most energetic X-rays emitted in each case when 40-keV electrons bombard (i) Mo target (Z = 42), (ii) Cu target (Z = 29) ? (e = 1.6 × 10^{-19} C).*

Solution : An *X*-ray photon of maximum energy is emitted where the electron loses the whole of its energy in a single collision with the target atom. Thus the maximum energy of the emitted *X*-ray is 40 keV, whichever be the target.

2. *What is the shortest wavelength of X rays produced in a tube when the applied voltage is 12.4 kV ?*

(h = 663 × 10^{-34} J-s, c = 3 × 10^{8} m/s and e = 1.6 × 10^{-10} C).

Solution : When electrons accelerated through a potential *V* strike a target, the maximum frequency ν_{max} (or minimum wavelength λ_{min}) of the emitted X-ray photon is given by

$$eV = h\nu_{max} = \frac{c}{\lambda_{min}}$$

so that $$\lambda_{min} = \frac{hc}{cV}.$$

Substituting $V - 12.4$ kV = 12400 volt and the given values of h, c and e, we get

$$\lambda_{min} = \frac{(6.63 \times 10^{-34} \text{ J-s})(3 \times 10^{8} \text{ m/s})}{(1.6 \times 10^{19} \text{ C})(12400 \text{ V})}$$

$$= 1.0 \times 10^{-10} \text{ m}$$

$$= 1.0 \text{ Å}.$$

3. An X-ray tube, operating at 66 kV, is producing continuous X-rays. Calculate the shortest wavelength limit of X-rays. (e = 1.6 × 10^−10 C. h = 6.6 × l0^−34 J-s, c = 3.0 × 10^8 m/s).

[Ans. 0.19Å]

4. What should be the kinetic energy of an electron so that its de Broglie wavelength is same as that of X-rays produced in an X-ray tube operating at 30 kV ? (h = 6.6 × l0^−34 J.s, c = 3.0 × 10^8 m/s, e =1.6 ×10^−9 C, m= 9.1 × 10^−31 kg).

Solution : The (minimum) wavelength of the X-rays produced is

$$\lambda_{min} = \frac{hc}{eV}$$

If K be the kinetic energy and p be the momentum of the electron of mass m, then we have

$$K = \frac{p^2}{2m}$$

But $$p = \frac{h}{\lambda},$$

where λ is the de Broglie wavelength of the electron.

$$\therefore \quad K = \frac{h^2}{2m\lambda^2}$$

Substituting the above value of λ, we get

$$K = \frac{h^2}{2mh^2c^2 / e^2V^2} = \frac{e^2V^2}{2mc^2}$$

$$= \frac{\left(1.6 \times 10^{-19} \text{ coul}\right)^2 \left(30{,}000 \text{ volt}\right)^2}{2\left(9.1 \times 10^{-31} \text{ kg}\right)\left(3.0 \times 10^8 \text{ ms}^{-1}\right)^2}$$

$$= 1.4 \times 10^{-16} \text{ joule.}$$

5. If the minimum wavelength recorded in the continuous X-ray spectrum of a 100-kV tube is 12.35×10^{-12} meter, calculate the value of Planck's constant. ($e = 1.6 \times 10^{-19}$ coulomb, $c = 3 \times 10^8$ m/s).

Solution : The minimum wavelength limit of the X-rays emitted from tube at V volt is given by

$$\lambda_{\text{min}} = \frac{hc}{eV}$$

so that $h = \dfrac{eV}{c} \lambda_{\text{min}}$

$$= \frac{\left(1.6\times 10^{-19} \text{ coul}\right)\left(100 \times 10^3 \text{ volt}\right)\left(12.35 \times 10^{-12} \text{ meter}\right)}{3 \times 10^8 \text{ m/s}}$$

$$= 6.59 \times 10^{-34} \text{ J-s.}$$

6. An X-ray tube produces X-rays of shortest wavelength of 0.25Å. Calculate the operating voltage. ($h = 6.6 \times 10^{-34}$ J-s, $e = 1.6 \times 10^{-19}$ C, $c = 3 \times 10^8$ m/s).

[Ans. 49.5 kV]

7. An X-ray machine produces X-rays of wavelengths 0.1 Å and greater. What voltage does it employ ?

[Ans. 1.24×10^5 volts]

8. 4n X-ray tube with Cu target is operated at 25 kV. The glancing angle for a NaCl crystal for the Cu K_α line is 15.8°.

Find the wavelength of this line. Also find the glancing angle for photons at the short wavelength limit.

(d for NaCl = 2.82 Å. h, c.

Solution: From Bragg's relation,

$$2d \sin \theta = n\lambda$$

For the first order, $n = 1$.

$$\therefore \quad \lambda = 2d \sin \theta$$

$$= 2 \times 2.82 \text{ Å} \times \sin 15.8°$$

$$= 2 \times 2.82 \text{ Å} \times 0.2723 = 1.53 \text{ Å}.$$

The wavelength of X-ray photons at the shortest limit is

$$\lambda_{\min} = \frac{hc}{eV} = 0.496 \text{ Å. (Solve)}$$

Let θ' *be* the glancing angle corresponding to $\lambda_{m/n}$. Then

$$\sin \theta' = \frac{0.496 \text{ Å}}{2 \times 1.82 \text{ Å}} = 0.088.$$

$$\therefore \quad \theta' = \sin^{-1} (0.088) = 5.1°.$$

9. *If the series limit of the Balmer series for hydrogen is 364.6 nm, calculate the atomic number of the element which gives X-ray wavelengths down to 0.1 nm.*

Solution : The Balmer series of hydrogen is given by

$$\frac{1}{\lambda} = R\left(\frac{1}{2^2} - \frac{1}{n^2}\right); \text{ n} = 3, 4, 5 \text{ ... ... } \infty.$$

For the series limit ($n = \infty$), we have

$$\frac{1}{\lambda_\infty} = \frac{R}{4}$$

$$\therefore \quad R = \frac{4}{\lambda\infty} = \frac{4}{364.6 \text{ nm}}.$$

The X-ray wavelengths of the *K*-series are given by

$$\frac{1}{\lambda} = R\,(Z-1)^2\left(\frac{1}{1^2}-\frac{1}{n^2}\right).$$

The minimum wavelength of this series occurs when $n = \infty$ and therefore

$$\frac{1}{\lambda} = R\,(Z-1)^2$$

or $$(Z-1)^z = \frac{1}{\lambda R}$$

Here $\lambda = 0.1$ nm and $R = 4/364.6$ nm. Thus

$$(Z-1)^2 = \frac{364.6\text{ nm}}{0.1\text{ nm}\times 4} = 911.5$$

or $$Z - 1 = 30.2.$$

$\therefore$ $Z = 31$, which corresponds to gallium.

10. How much energy goes into the quantum emitted when an L-electron (i.e. an electron from the n = 2 state) drops into the empty K-state in copper (Z = 29) ?

Given R = 109737 cm^{-1} ; l cm^{-1} = 1.239 × 10^{-4} eV.

Solution : The X-ray wavelength λ emitted, when an $n = 2$ electron drops to the $n = 1$ state, is given by

$$\frac{1}{\lambda} = R\,(Z-b)^2\left(\frac{1}{1^2}-\frac{1}{2^2}\right).$$

For the *K*–line, $b = 1$. Therefore

$$\frac{1}{\lambda} = \frac{3}{4}R\,(Z-1)^2.$$

Putting the given values of *R* and Z :

$$\frac{1}{\lambda} = \frac{3}{4}\,(109737\text{ cm}^{-1})\,(29-1)^2$$

$$= 6.4525\times 10^7\text{ cm}^{-1}.$$

The corresponding energy emitted is

$(6.4525 \times 10^7) \times (1.239 \times 10^{-4}) = 7994.6$ eV.

11. Estimate the value of the wavelength of K_α line of silver (Z = 47). Given R = 109737 cm^{-1}.

Solution : From Moseley's law, the wavelength of K_α-line is given by

$$\frac{1}{\lambda_{K\alpha}} = R\,(Z-1)^2.\left(\frac{1}{1^2}-\frac{1}{2^2}\right).$$

$$= \frac{3}{4}\,R\,(Z-1)^2$$

$$= \frac{3}{4} \times 109737 \text{ cm}^{-1} \times (46)^2$$

$$= 1.74 \times 10^8 \text{ cm}^{-1}.$$

$$\therefore \qquad \lambda = \frac{1}{1.74 \times 10^8} \text{ cm} = \frac{10^8}{1.74 \times 10^8} \text{ Å}$$

$$= 0.57 \text{ Å}.$$

12. What element has a K_α X-ray line of wavelength 1.785A? R = 109737 cm^{-1}.

[**Ans.** Z = 27 (Co)]

13. If the K_α–radiation of Mo (Z = 42) has a wavelength of 0.71 Å, calculate the wavelength of the corresponding radiation of Cu (Z = 29).

Solution : From Moseley's law; we have

$$\frac{1}{\lambda_{K\alpha}} \propto (Z-1)^2.$$

$$\therefore \qquad \frac{\lambda_{cu}}{\lambda_{mo}} = \frac{(Z_{mo}-1)^2}{(Z_{cu}-1)^2} = \frac{(41)^2}{(28)^2}$$

But $\lambda_{mo} = 0.71$ Å.

$$\therefore \qquad \lambda c_u = (0.71 \text{ Å}) \times \frac{(41)^2}{(28)^2} = 1.52 \text{ Å}.$$

14. If the K–, L–, M-energy levels of platinum involved in the emission of the K-series of X-rays of this element are 78000, 12000 and 3000 eV respectively, calculate the wavelengths of the K_α and K_β lines of platinum. ($h = 6.63 \times 10^{-34}$ joule-sec, $c = 3.0 \times 10^8$ meter/sec, 1 eV = 1.6×10^{-19} joule).

Solution : The K_α–line is emitted when an electron drops from the *L*-shell to the *K*-shell. Hence the energy of a K_α-photon is the difference in energies of these two shells, *i.e.*

$$h\nu_{K\alpha} = 78000 - 12000 = 66000 \text{ eV}$$

$$= 66000 \times (1.6 \times 10^{-19}) \text{ joule.}$$

$$\therefore \quad \upsilon_{k\alpha} = \frac{6600 \times \left(1.6 \times 10^{19}\right)}{6.63 \times 10^{-34}} \text{ sec}^{-1}.$$

The corresponding wavelength is

$$\lambda_{K\alpha} = \frac{c}{\upsilon_{K\alpha}}$$

$$= \frac{\left(3.0 \times 10^{8}\right) \times \left(6.63 \times 10^{-34}\right)}{66000 \times \left(1.6 \times 10^{-19}\right)}$$

$$= 1.88 \times 10^{-11} \text{ meter}$$

$$= 0.188 \text{ Å}.$$

The K_β-line is emitted when an electron drops from the *M*-shell to the *K*-shell. Thus

$$h\upsilon_{K\beta} = 78000 - 3000 = 75000 \text{ eV}.$$

$$\therefore \quad \lambda_{K\beta} = \frac{\left(3.0 \times 10^{8}\right) \times \left(6.63 \times 10^{-34}\right)}{75000 \times \left(1.6 \times 10^{-19}\right)}$$

$$= 1.65 \times 10^{-11} \text{ meter} = 0.165\text{Å}.$$

15. If the K-absorption limit of Uranium is 0.107 Å, find the minimum potential difference required across an X-ray tube to excite the K-series ?

Solution : If v_K be the frequency of the K-absorption limit for the target material, the critical voltage V_K required is given by

$$eV_K = hv = \frac{hc}{\lambda_K}$$

$$\therefore \qquad Vk = \frac{hc}{e\lambda_K}$$

Now $$h = 6.63 \times 10^{-34} \text{ joule-sec},$$

$$c = 3.0 \times 10^{8} \text{ meter/sec},$$

$$e = 1.6 \times 10^{-19} \text{ coulomb}$$

and $$\lambda_K = 0.107 \text{ Å} = 0.107 \times 10^{-10} \text{ meter.}$$

$$\therefore \qquad V_K = \frac{\left(6.63 \times 10^{-34}\right) \times \left(3.0 \times 10^{8}\right)}{\left(1.6 \times 10^{-19}\right) \times \left(0.107 \times 10^{-10}\right)}$$

$$= 116 \times 10^{3} \text{ volts} = 116 \text{ kV.}$$

3

Electromagnetic Waves

X-rays

Roentgen, in 1895, discovered that when fast-moving electrons strike a dense target, a highly-penetrating radiation is produced. He called this radiation as *X*-rays. Later experiments showed that *X*-rays are electromagnetic waves having a wavelength of the order of 1 Å.

Production of X-rays

Coolidge Tube: An *X*-ray tube contains a cathode which serves as the source of electrons and an anode (or target) upon which the electrons strike. A modern tube designed by Coolidge consists of a glass bulb evacuated to about 10^{-6} cm of Hg so that no discharge can pass in it. The cathode is a tungsten filament *F* heated by a low-voltage current drawn from the separate tapping of a transformer.

A cylindrical metal shield *S* surrounds the filament and is kept at a negative potential with respect to the filament. It focuses the electrons emitted from the filament on a tungsten target T. The target is mounted on the surface of a copper block fixed to the end of a copper rod. The rod is fitted with fins for air-cooling, *or* made hollow and cooled by circulating water through it.

A p.d. of about 20,000 volts is applied across the tube by means of a step-up transformer. The electrons emitted from the filament are accelerated and focussed on the target which radiate X-rays. The intensity of X-rays depends on the rate of emission of electrons from the cathode, and can be controlled by varying the filament-current by means of a rheostat R in the filament circuit. The quality (penetrating power) of X-rays depends on the p.d. across the tube, and can be controlled *independently* of the intensity. The electrons emitted from the filament strikes the target only when the target is at a positive potential with respect to the filament. The tube thus acts as its own rectifier, running easily on A.C. mains.

Properties of X-rays: The X-rays show the following important properties :

(*i*) X-rays travel in straight lines with the speed of light.

(*ii*) They blacken a photographic plate.

(*iii*) They are not deflected by either electric or magnetic fields.

(*iv*) They liberate photoelectrons from certain metals on which they fall.

(*v*) They ionise the gas through which they pass.

(*vi*) They produce fluorescence on screens of zinc sulphide, barium platinocyanide, calcium tungsten, etc.

(*vii*) They penetrate considerable thickness of substances which are opaque to ordinary light such as wood, paper, cardboard, thin sheets of metals, flesh, etc. The penetration decreases as the density of the substance increases. X-rays cannot penetrate through heavy metals and bones, and cast shadow of these objects when placed in their path.

(*viii*) They exhibit reflection, refraction, interference, diffraction and also polarisation under special circumstances.

(*ix*) They produce a reddening of the skin which may prove fatal.

(*x*) When a beam of primary *X*-rays falls on a plate of some chosen element, a part of the radiation goes through the plate and the rest is transformed into heat, or into some secondary radiation. The latter consists of scattered *X*-rays, characteristic *X*-rays, scattered β-rays and characteristic β-rays (Fig.).

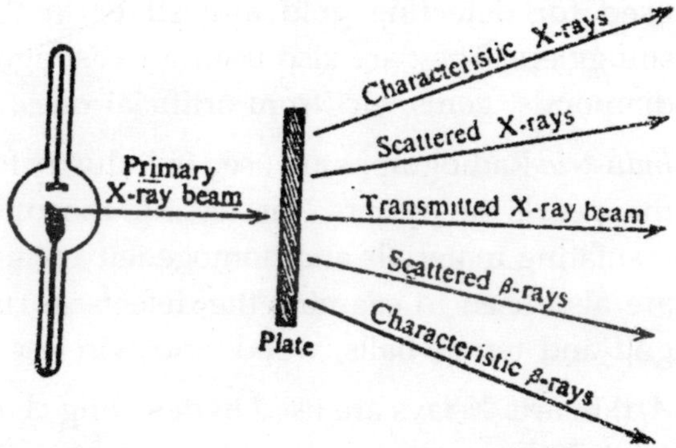

Detection of X-rays: *X*-rays are detected by means of a fluorescent screen, special photographic film, an ionisation chamber or an electroscope. The last two are often used to measure the number of ions formed in a gas by *X*-rays which is directly proportional to the intensity of *X*-rays.

Uses of X-rays: *X*-rays have been put to a large number of uses :

(i) *Surgery: X*-ray photographs, called radiographs, are used to detect fractures of bones, tuberculosis of lungs, foreign matter like bullets, formation of bones and stones in the body and observing the progress of a healing bone.

(ii) *Radio-therapy:* Controlled X-ray exposures are used to cure intractable skin diseases and malignant growths. Soft X-rays are used if the affected parts are superficial while hard X-rays are used for deep-seated organs.

(iii) *Engineering:* Radiographs are used in engineering for detecting faults, cracks, flaws and gas pockets in finished metal products. They are used to test weldings, castings and moulds.

(iv) *Detective Department:* X-rays are used for detection of explosives, opium and other contraband goods in sealed parcels and leather cases. At custom ports they are used for detecting gold and silver in the body of smugglers. They are also used for distinguishing real diamonds, gems, etc. from artificial ones.

(v) *Industry:* Radiographs are used in industry for detecting the pearls in oysters, for testing the uniformity of insulating materials and homogeneity of timber. They are also used to examine the defects in rubber tyres, golf and tennis balls, wood, and wireless valves.

(vi) *Art:* In arts X-rays are used in detecting changes in old oil paintings.

(vii) *Scientific Research:* X-rays are used in investigating the structure of crystals, constitution and properties of atoms and arrangement of atoms and molecules in complex substances.

Wave-like Character of X-rays: The observed properties of X-rays could, however, not establish whether X-rays were waves or streams of uncharged particles. Barkla, in 1906, showed by a double scattering experiment that X-rays could be polarised like visible light waves and thus established that they are transverse electromagnetic waves.

The plan of Barkla's experiment is shown in Fig below. A beam of *X*-rays moving along the *z*-axis is incident on a carbon block *A*. These are scattered by carbon atoms. If the incident beam is a (transverse) electromagnetic wave, it has electric vectors in the *xy* plane only (because electric vector is always perpendicular to the direction of propagation of the wave). These electric vectors cause the electrons in the carbon atoms to vibrate in the *xy* plane which re-radiate in the form of scattered *X*-rays.

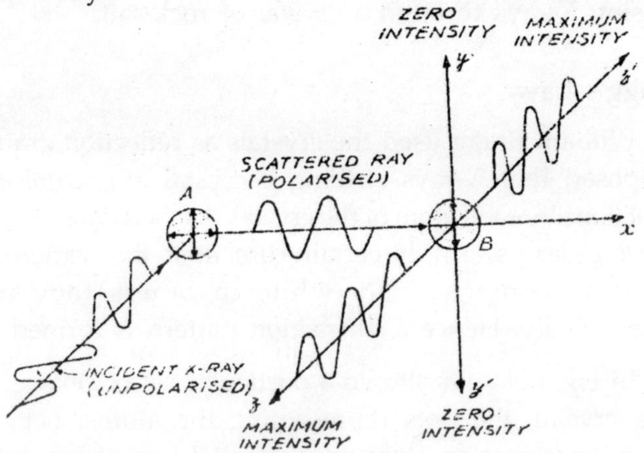

Now, a scattered *X*-ray moving along the *x*-axis can have an electric vector along the *y*-axis only and thus it is plane-polarised. To show that the rays along the *x*-axis are plane-polarised, Barkla scattered them again from a second carbon block *B*. The electrons in this block vibrate along the *y*-axis only. Hence they re-radiate *X*-rays exclusively in the *x*-*z* plane, and not at all along the *y*-axis. Barkala found that there were no *X*-rays along the *y*-axis. This established the transverse wave character of X-rays.

If *X*-rays were a stream of particles or longitudinal waves, the above observation could by no means be understood.

Crystal as a Three-dimensional Grating for X-rays: Waves show diffraction effects when they are broken up by regularly-spaced scattering points with separation of the order of the wavelength. X-rays are electromagnetic waves of the order of 1 Å. Such short waves cannot be diffracted by optical gratings which are too coarse for them.

Laue, in 1912, pointed out that crystals have geometrically-regular, periodic arrangements of atoms in space, with separations comparable to the X-ray wavelengths. Hence crystals could act as natural three-dimensional grating for X-rays. Laue's students actually obtained a diffraction pattern by passing X-rays through a crystal of rock-salt.

Bragg's Law

William Bragg used the crystals as reflection gratings. He supposed that X-rays striking a crystal are scattered in all directions by each atom of the crystal. As the atoms are arranged in a regular pattern, in certain directions the scattered waves interfere constructively while in others they interfere destructively. Hence a diffraction pattern is formed.

In Fig. below is shown a particular set of cleavage planes in a crystal, the dots representing the atoms. Let d be the interplanar spacing. Suppose a parallel X-ray beam is incident at glancing angle θ. It is scattered by the atoms A and B in random directions. Constructive interference takes place only between those scattered waves which are parallel and have a path difference of $n\lambda$, where λ is the X-ray wavelength and n is an integer. *It is seen that for such rays the scattering angle is equal to the glancing angle* θ. The path difference is

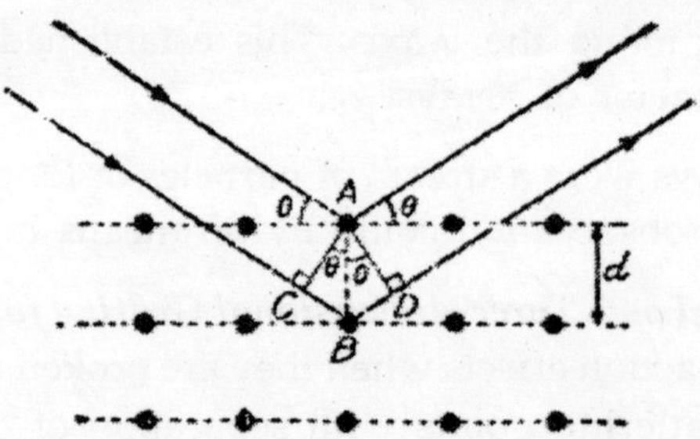

$$CB + BD = d \sin \theta + d \sin \theta$$

$$= 2d \sin \theta.$$

For constructive interference, we must have

$$2d \sin \theta = n\lambda \qquad n = 1, 2, 3................$$

This is Bragg's equation. It shows that for given values of λ and d, there are a number of permitted directions defined by θ (for $n = 1, 2, 3,......$) in which mirror-like reflection will occur. The images obtained in these directions are the first, second, third,..... order diffraction images.

If the incident X-ray beam consists of several wavelengths, a number of spectra are obtained.

Bragg's X-ray Spectrometer: It is shown in Fig. X-rays from a tube pass through fine slits S_1 and S_2 which collimate in into a fine pencil. This pencil falls on a crystal C mounted on a circular table capable of rotation about a vertical axis. The position of the table can be read on a graduated scale with a vernier V_1. The reflected beam after passing through the slit S_3 enters the ionisation chamber D, filled with methyl iodide which absorbs X-rays strongly. The chamber is mounted on an arm which can be rotated about the same axis as the crystal. Its position can be read by a second vernier V_2. The gas in the chamber is ionised by the X-rays. The resulting ionising current, which is measured by the electrometer E, is a measure of the intensity of X-rays reflected by the crystal.

Working: The crystal C is placed on the table such that the face from which the X-rays are to be reflected contains the axis of rotation. The crystal is rotated through small angles, and also the ionisation chamber through double the angles, and the ionisation current is measured each time. It is found that as the glancing angle θ is varied, the ionisation in the chamber also varies. Maximum ionisation occurs for values of θ given by Bragg's law.

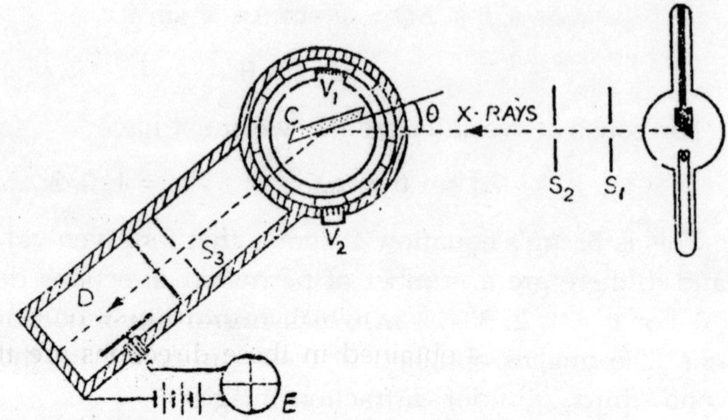

A graph between ionisation current and the glancing angle θ is shown in Fig. below for three wavelengths. It is seen that the intensity of *X*-rays decreases for higher orders while the difference between the glancing angles for the three wavelengths increases as the order of the spectrum increases. The spectral lines are superimposed on a continuous spectrum represented by the dotted line.

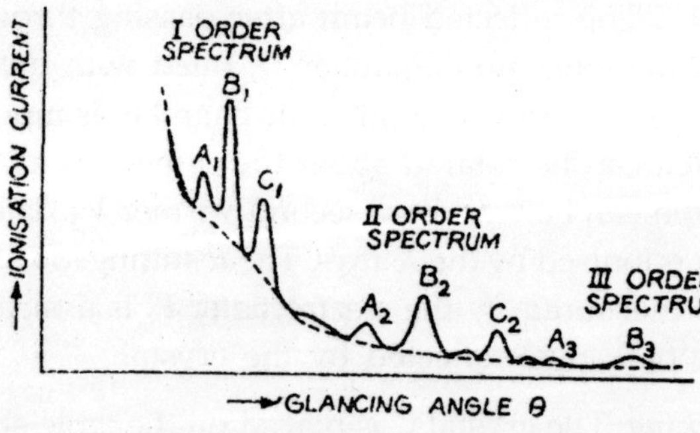

Determination of Wavelength

For a particular wavelength λ, we have

$$2d \sin \theta = n \lambda$$

or

$$\frac{\lambda}{d} = \frac{2 \sin \theta}{n}$$

By observing those values of θ at various orders n for which sharp reflection occurs, a mean value of λ/d is determined. Now, if the spacing of the crystal planes, d, is known, the wavelength λ of X-rays can be calculated.

As an example, it has been found from the X-ray study of NaCl crystal that the spacing between the (100) planes of this crystal is 2.82 Å. An X-ray line from a palladium target gives a first-order reflection at a glancing angle of 6° from the (100) planes of NaCl. Then, from the Bragg's equation, we have

$$\lambda = \frac{2d \sin \theta}{n}$$

$$= \frac{2 \times 2.82 \text{ Å} \times \sin 6^\circ}{1} = 0.59 \text{ Å}$$

The accuracy of the values of λ so obtained, however, depends upon of the accuracy with which the value of crystal spacing d is known.

In 1925, Compton and Doan developed the technique to obtain the diffraction of X-rays by ruled grating. Hence now we can determine independently the wavelength of X-rays with precision.

PROBLEMS

1. An X-ray beam has a wavelength of 1 Å. Calculate (i) energy of the X-rays photon, (ii) momentum of the photon, (iii) mass of the photon, (iv) rest mass of the photon, ($h = 6.6 \times 10^{-14}$ J-s, $c = 3.0 \times 10^8$ m/s).

Solution :

(*i*) The energy of the X-rays photon is given by

$$E = h\upsilon = \frac{hc}{\lambda}$$

$$= \frac{\left(6.6 \times 10^{-34}\,\text{J-s}\right)\left(3.0 \times 10^{8}\,\text{ms}^{-1}\right)}{\left(1 \times 10^{-10}\,\text{m}\right)}$$

$$= 1.98 \times 10^{-15}\ \text{J}.$$

(*ii*) The momentum of the photon is given by

$$p = \frac{h}{\lambda}$$

$$= \frac{6.6 \times 10^{-34}\,\text{J-s}}{1 \times 10^{-10}\,\text{m}}$$

$$= 6.6 \times 10^{-24}\ \text{N.s.}$$

(*iii*) The (kinetic) mass of the photon is given by

$$m = \frac{E}{c^2} \qquad [\because E = mc^2]$$

$$= \frac{1.98 \times 10^{-15}\,\text{J}}{\left(3.0 \times 10^{8}\,\text{ms}^{-1}\right)^2}$$

$$= 2.2 \times 10^{-22}\ \text{kg}$$

(*iv*) The rest mass of a photon is zero.

2. *An X-ray tube is operated at 20 kV. Let's find the maximum speed of the electrons striking the anticathode. The mass and the charge of the electron are 9.1 × 10^{-31} kg and 1.6 × 10^{-19} coulomb respectively.*

Solution : When an electron of charge e coulomb is accelerated through a potential difference of V volt, it acquires a maximum energy of eV joule. If m kg be the mass and v_{max}, meter/sec the maximum speed of the electron, then we have

$$\frac{1}{2} m v_{max}{}^2 = eV$$

$$v_{max} = \sqrt{\left(\frac{2eV}{m}\right)}$$

$$= \sqrt{\left(\frac{2 \times \left(1.6\ 10^{-19}\,\text{coul}\right) \times (20000\ \text{volt})}{9.1 \times 10^{-31}\,\text{kg}}\right)}$$

$$= 8.4 \times 10^{7}\ \text{m/s.}$$

3. Using a Bragg's X-ray spectrometer the glancing angle for the first-order spectrum was observed to be 8°. The wavelength of the X-rays, if d = 2.82 × 10⁻⁸ cm may be calculated.

Solution : The Bragg's equation for X-ray reflection is

$$2d \sin\theta = n\lambda,$$

where d is the interplanar distance and n is the order. From this, we have

$$\lambda = \frac{2d \sin\theta}{n}$$

Here $n = 1$, $\theta = 8°$ and $d = 2.82 \times 10^{-8}$cm $= 2.82$ Å

$$\therefore \qquad \lambda = \frac{2 \times 2.82\ \text{Å} \times \sin 8°}{1}$$

$$= 2 \times 2.82\ \text{Å} \times 0.139 = 0.785\ \text{Å}.$$

4. The line A of an X-ray beam gives a first-order reflection maximum at a glancing angle of 30° to the smooth face of a crystal. Line B (λ = 0.97 Å) gives a third-order reflection maximum at an angle of 60° from the same face of the same crystal. We can find the wavelength of the line-A.

The Bragg's equation is $2d \sin\theta = n\lambda$. From this,

we have

$$2d \sin 30° = \lambda_A$$

and $$2d \sin 60° = 3\lambda_B = 3 \times 0.97\ \text{Å}$$

$$\therefore \qquad \lambda_A = \frac{3 \times 0.97\ \text{Å}}{\sin 60°} \times \sin 30° = 1.68\text{Å}$$

5. In Bragg's spectrometer the first-order spectrum for X-rays of wavelength 0.36 Å is obtained at an angle of 4.8°. The effective spacing between the atomic planes may be found.

Solution : The Bragg's equation is

$$2d \sin\theta = n\lambda, \qquad \text{where } n = 1, 2, 3, ...$$

$$\therefore \qquad d = \frac{n\lambda}{2\sin\theta}$$

Here $n = 1$, $\lambda = 0.36$ Å and $\sin\theta = \sin 4.8° = 0.0837$.

$$\therefore \qquad d = \frac{1 \times 0.36 \text{ Å}}{2 \times 0.0837} = 2.15 \text{ Å}.$$

6. Diffraction of X-rays in a crystal gives first peak at a glancing angle of 11.8°. If the wavelength of X-rays used be 1.147 Å, calculate the spacing between the atomic layers of the crystal.

Solution : The answer is 2.80 Å.

7. A set of lattice planes reflects *X*-rays of wavelength 1.32 Å at a glancing angle of 9° 30'. The possible spacings of this set of planes may be deduced.

Solution : The Bragg's equation is

$$2d \sin\theta = n\lambda, \qquad \text{where } n = 1, 2, 3, ...$$

$$\therefore \qquad d = \frac{n\lambda}{2\sin\theta}$$

Here $\lambda = 1.32$ Å, $\sin\theta = \sin(9° 30') = 0.165$.

$$\therefore \qquad d = \frac{n \times 1.32 \text{ Å}}{2 \times 0.165} = 4n\text{Å} = 4, 8, 12, ... \text{ Å}.$$

8. Let's see as to at what angles may a ray of wavelength 0.440 Å be reflected from the cube face of a rock-salt crystal (d = 2.814 Å).

Solution : The Bragg's condition for *X*-ray reflection is

$$2d \sin \theta = n\lambda,$$

where d is the interplanar distance and n is the order. From this, we have

$$\sin \theta = \frac{n\lambda}{2d} = \frac{n\left(0.440\ \text{Å}\right)}{2 \times 2.814\ \text{Å}} = 0.07818\ n.$$

For first, second, ... order, (n = 1, 2.......), we have

$$\sin \theta = 0.07818,\ 0.15636,\ ...$$

or $$\theta = 4.5°,\ 9°,\ ...$$

9. The glancing angle on the cube face [100] of a rock-salt crystal (d = 2.814 Å) corresponding to the second-order reflection for X-rays of λ = 0.710 Å may be calculated.

Solution : The answer is 14.6°.

10. The distance between adjacent atomic planes in calcite is 3Å. Let's find out the smallest angle between these planes and an incident beam of 0.3 Å. X-rays at which these X-rays can be detected.

Hint : For smallest glancing angle, n = 1. and

Ans. $\sin^{-1} 0.05$.

11. The spacing between the successive planes in NaCl crystal is 2.82 Å. X-rays incident on this crystal give rise to first-order reflection at grazing incidence of 8° 35′. We may calculate the wavelength of X-rays and the angle for second-order and third-order reflections (sin 8° 35′ = 0.1492). The longest wavelength that can be analysed by the crystal is calculated here.

Solution : The Bragg's equation is

$$2d \sin \theta = n\lambda \qquad ...\ (i)$$

so that $$\lambda = \frac{2d \sin \theta}{n}$$

Here d = 2.82 Å, sin θ = sin 8° 35′ = 0.1492 and n = 1.

$$\therefore \qquad \lambda = 2 \times (2.82 \text{ Å}) \times 0.1492 = 0.84 \text{ Å}.$$

Now from eq. (*i*), we have

$$\sin \theta = \frac{n\lambda}{2d}.$$

For the second order (n = 2), we get

$$\sin \theta = \frac{2 \times 0.84}{2 \times 2.82} = 0.2979.$$

$$\theta = 17° \ 18'.$$

Similarly, for the third order (n = 3), we get

$$\sin \theta = \frac{3 \times 0.84}{2 \times 2.82} = 0.4468$$

$$\therefore \qquad \theta = 26°36'.$$

For the longest wavelength θ = 90° and n = 1.

$$\therefore \qquad \lambda = 2d = 2 \times 2.82 = 5.64 \text{ Å}.$$

12. ***The result of the longest wavelength that can be analysed by a rock-salt crystal of spacing 282 Å in the second order is 2.82Å.***

4

Hydrogen Atom

Schrodinger Equation of Hydrogen Atom

The hydrogen atom consists of a single electron of mass m and charge - e, moving about a nucleus of mass M and charge + e in a potential field $V(r)$ which corresponds to Coulombian attraction $\frac{e^2}{4\pi\epsilon_0 r^2}$ between the electron and the nucleus. In general, the nucleus is not stationary. Its motion can, however, be taken into account by treating the system as a single particle of (reduced) mass μ moving in the potential field $V(r)$, where $\mu = mM/(m+M)$. The Schrodinger equation for the motion is

$$\nabla^2 \psi + \frac{8\pi^2\mu}{h^2} [E - V(r)] \psi = 0.$$

Converting the Laplacian operator ∇^2 in spherical coordinates, we get

$$\frac{1}{r^2}\frac{\partial}{\partial r}\left(r^2\frac{\partial \psi}{\partial r}\right)+\frac{1}{r^2 \sin\theta}\frac{\partial}{\partial \theta}\left(\sin\theta\frac{\partial \psi}{\partial \theta}\right)+\frac{1}{r^2 \sin^2\theta}\frac{\partial^2 \phi}{\partial \phi^2}$$

$$+\frac{8\pi^2\mu}{h^2}[E-V(r)]\,\psi=0. \quad \ldots (i)$$

This is the Schrodinger equation of hydrogen atom in spherical coordinates. To solve this equation we must separate the radial (r) and angular (θ, ϕ) variables.

Separation of Variables

Let us substitute

$$\psi(r, \theta, \phi) = R(r)\,Y(\theta, \phi)$$

in eq. (i). We get

$$\frac{1}{r^2}\frac{d}{dr}\left(r^2\frac{dR}{dr}\right)Y+\frac{1}{r^2 \sin\theta}\frac{\partial}{\partial \theta}\left(\sin\theta\frac{\partial Y}{\partial \theta}\right)R$$

$$+\frac{1}{r^2\sin^2\theta}\frac{\partial^2 Y}{\partial \phi^2}R+\frac{8\pi^2\mu}{h^2}[E-V(r)]\,RY=0.$$

Multiplying the entire equation by $\frac{r^2}{RY}$ and rearranging, we get

$$\frac{1}{R}\frac{d}{dr}\left(r^2\frac{dR}{dr}\right)+\frac{8\pi^2\mu r^2}{h^2}[E-V(r)]$$

$$=-\frac{1}{Y}\left[\frac{1}{\sin\theta}\frac{\partial}{\partial \theta}\left(\sin\theta\frac{\partial Y}{\partial \theta}\right)+\frac{1}{\sin^2\theta}\frac{\partial^2 Y}{\partial \phi^2}\right].$$

The left-hand side of this equation is a function of R only, while the right-hand side is a function of Y only. Hence each side must be equal to the same constant. Let this constant be $l(l+1)$. Thus we get a radial equation

$$\frac{1}{R}\frac{d}{dr}\left(r^2\frac{dR}{dr}\right)+\frac{8\pi^2\mu r^2}{h^2}[E-V(r)]=l(l+1)$$

or $$\frac{1}{r^2}\frac{d}{dr}\left(r^2\frac{dR}{dr}\right)+\left[\frac{8\pi^2\mu}{h^2}\{E-V(r)\}-\frac{l(l+1)}{r^2}\right]R=0, \quad \text{... (ii)}$$

and an angular equation

$$-\frac{1}{Y}\left[\frac{1}{\sin\theta}\frac{\partial}{\partial\theta}\left(\sin\theta\frac{\partial Y}{\partial\theta}\right)+\frac{1}{\sin^2\theta}\frac{\partial^2 Y}{\partial\phi^2}\right]=l\ (l+1). \quad \text{... (iii)}$$

The later equation can be further splitted by substituting $Y(\theta, \phi) = \Theta(\theta)\ \Phi(\phi)$.

Thus we get

$$-\frac{1}{\Theta\,\Phi}\left[\frac{1}{\sin\theta}\frac{d}{d\theta}\left(\sin\theta\frac{d\Theta}{d\theta}\right)\Phi+\frac{1}{\sin^2\theta}\frac{d^2\Phi}{d\phi^2}\Theta\right]=l\ (l+1).$$

Multiplying the entire equation by $\sin^2 9$ and rearranging, we get

$$\frac{1}{\Theta}\sin\theta\frac{d}{d\theta}\left(\sin\theta\frac{d\Theta}{d\theta}\right)+l(l+1)\sin^2\theta = \frac{1}{\Phi}\frac{d^2\Phi}{d\phi^2}.$$

Again, the two sides of this equation are functions of different variables and so each must be equal to the same constant mt^2 (say). Thus we get

$$\frac{1}{\Theta}\sin\theta\frac{d}{d\theta}\left(\sin\theta\frac{d\Theta}{d\theta}\right)+l(l+1)\sin^2\theta = m_l^2$$

or $$\frac{1}{\sin\theta}\frac{d}{d\theta}\left(\sin\theta\frac{d\Theta}{d\theta}\right)+\left[l(l+1)-\frac{m_l^2}{\sin^2\theta}\right]\Theta=0, \quad \text{... (iv)}$$

and $$-\frac{1}{\Phi}\frac{d^2\Phi}{d\phi^2} = mi^2$$

or $$\frac{d^2\Phi}{d\phi^2}+ml^2\Phi=0. \quad \text{... (v)}$$

Thus we have broken the Schrodinger equation into three differential equations (*ii*), (*iv*) and (*v*) which are respectively in R alone, in Θ alone and in Φ alone. When we try to find "acceptable' solutions of those equations, the quantum numbers automatically come in.

Wave Functions of Hydrogen Atom

Entrance of Quantum Numbers: The solution of eq. (*v*) is

$$\Phi_{ml}\,(\phi) = Ae^{\,iml\,\phi}, \qquad \text{... }(vi)$$

where A is integration constant. It can be shown that $\Phi_{ml}\,(\phi)$ is a single-valued function of position only when ml is a positive or negative integer or zero, that is

$$mi = 0, \pm 1, \pm 2, \pm 3, \ldots \ldots$$

ml is known as 'magnetic quantum number'.

The solution of eq. (iv) is known to be

$$\Theta_{l,\,ml}\,(\theta) = N_{l,\,ml}\;p_l^{|ml|}\,(\cos\theta), \qquad \text{... }(vii)$$

where Nl, ml is constant and $Pi^{|mi|}$ is 'associated Legendre polynomial' of degree l and order m_l. The solution is found to be acceptable only when l is a positive integer, that is,

$$l = 0, 1, 2, \ldots\ldots\ldots$$

l is known as 'orbital quantum number'.

Further the functions $\Theta._{l,mi}\,(\theta)$ vanish if $|ml|$ exceeds l. This means that m_l can take values only upto $\pm\, l$.

For solving the radial equation (*ii*) we must specify $V\,(r)$.

In the present case it is $-\dfrac{e^2}{4\pi \in_0 r}$. Then the equation is

$$\frac{1}{r^2}\frac{d}{dr}\left(r^2\frac{dR}{dr}\right) + \left[\frac{8\pi^2\mu}{h^2}\left\{E + \frac{e^2}{4\pi \in_0 r}\right\} - \frac{l(l+1)}{r^2}\right] R = 0$$

or $$\frac{d^2R}{dr^2}+\frac{2}{r}\frac{dR}{dr}+\left[\frac{8\pi^2\mu}{h^2}\left\{E+\frac{e^2}{4\pi\in_0 r}\right\}-\frac{l(l+1)}{r^2}\right]R=0.$$

The solution of this equation is known to be

$$R_{n,l}(r)=N_{n,l}\left(\frac{2r}{na_0}\right)^i e^{-r/na_0}\,L_{n+l}^{2l+1}\left(\frac{2r}{na_0}\right), \quad \text{... (viii)}$$

where $N_{n,l}$ is constant and L_{n+l}^{2l+1} is 'associated Legen poly-nomial' of degree $n = l - 1$ and order $2l + 1$. The parameter a_0 is

$$a_0=\frac{\in_o h^2}{\pi\mu e^2}$$

is the radius of first Bohr orbit. The solution is found to be acceptable only when $n — l—1$ is a positive integer. This means that n, I and m_l are restricted to the following values:

$$n = 1, 2, 3,.........\infty$$

$$l = 0, 1,2,.........(n\text{–}l)$$

$$ml = 0, \pm 1, \pm 2,.........\pm 1$$

n is known as 'total quantum number'.

Thus the wave functions of hydrogen atom are given by

$$\psi_{n,l,ml}(r,\theta,\phi) = R_{n'l}(r)\,{}^{\Theta}\,l_{,ml}(\theta)\,\Phi_{ml}(\phi)$$

where R, Θ and Φ are given by eq. (*viii*), (*vii*), and (*vi*) respectively. The real form of the wave functions of hydrogen atom for the ground state ($n = 1, l = 0.\ m_l = 0$) is given below:

$R_{1,0}(r)$	$\Theta_{0,0}(\theta)$	$\Phi_0(\phi)$	$\psi_{1,0,0}(r,\theta,\phi)$
$\frac{2}{a_0^{3/2}}e^{-r/a_0}$	$\frac{1}{\sqrt{2}}$	$\frac{1}{\sqrt{(2\pi)}}$	$\frac{1}{\sqrt{(\pi a_0^3)}}e^{-r/a_0}$

Wave Mechanical Interpretation of n, I, ml :— The radial wave equation for the hydrogen atom can be solved only when

E has a continuous range of positive values, or a discrete set of *negative* values (which correspond to *bound* states), given by

$$E = \frac{\mu e^4}{8 \epsilon_0^2 h^2}\left(\frac{1}{n^2}\right), \qquad n = 1,2,3,...$$

These are same as obtained by old quantum theory. Thus *n* describes the quantisation of electron energy in the hydrogen atom.

Let us now find the interpretation of *l* and m_l. The wave mechanical operators for the square of the angular momentum and for the z-component of angular momentum are

$$L_2 = -\frac{h^2}{4\pi^2}\left[\frac{1}{\sin\theta}\frac{\partial}{\partial\theta}\left(\sin\theta\frac{\partial}{\partial\theta}\right) + \frac{1}{\sin^2\theta}\frac{\partial^2}{\partial\phi^2}\right] \qquad ...(ix)$$

and $$L_z = -\frac{ih}{2\pi}\frac{\partial}{\partial\phi}. \qquad ...(x)$$

The angular eq. (*iii*) can be written as

$$\left[\frac{1}{\sin\theta}\frac{\partial}{\partial\theta}\left(\sin\theta\frac{\partial}{\partial\theta}\right) + \frac{1}{\sin^2\theta}\frac{\partial^2}{\partial\phi^2}\right] Y_{1,mi}(\theta,\phi)$$

$$= -l(l+1)\, Y_{l,mi}(\theta,\phi).$$

Multiplying both sides by $-\frac{h^2}{4\pi^2}$, we get

$$-\frac{h^2}{4\pi^2}\left[\frac{1}{\sin\theta}\frac{\partial}{\partial\theta}\left(\sin\theta\frac{\partial}{\partial\theta}\right) + \frac{1}{\sin^2\theta}\frac{\partial^2}{\partial\phi^2}\right] Y_{l,ml}(1,\phi)$$

Substituting eq. (ix) in it, we have

$$L^2\, Y_{l,ml}(\theta,\phi) = l(l+1)\frac{h^2}{4\pi^2}\, Y_{l,ml}(\theta,\phi).$$

Thus the angular wave functions $Y,_{l,ml}(\theta,\phi)$ of hydrogen atom are also the wave functions of the square of the angular momentum operator with eigenvalues $l(1+1)\frac{h^2}{4\pi^2}$. This means that the electron in the hydrogen atom has an angular momentum given by

$$L = \sqrt{[l(l+1)]}\frac{h}{2\pi},$$

where $l = 0, 1, 2,...(n - 1)$. Thus, like energy, the angular momentum is also quantised, and this quantisation is described by the orbital quantum number l.

Since the magnitude of the electron angular momentum is quantised, its direction is also expected to be quantised with respect to an external magnetic field (space quantisation). Let us consider the Φ-equation (*vi*) :

$$\Phi_{ml}(\phi) = Ae^{imt\phi}$$

Differentiating it, we get

$$\frac{d\Phi_{mt}(\phi)}{d\phi} = Ae^{iml\phi}(i\ mi)$$

$$= i\ mi\ \Phi_{ml}(\phi).$$

Multiplying both sides by $-\frac{ih}{2\pi}$, we get

$$-\frac{ih}{2\pi}\frac{d\Phi_{ml}(\phi)}{d\phi} = \frac{m_l h}{2\pi}\Phi_{ml}(\phi).$$

Substituting eq, (*x*) in it, we have

$$L_z\Phi_{ml}(\phi) = \frac{m_l h}{2\pi}\Phi_{ml}(\phi).$$

Thus the wave function $\Phi_{ml}(\phi)$ of hydrogen is also the wave function of z-component of the angular momentum operator with eigenvalues $\frac{m_l h}{2\pi}$. This means that the component of the electron angular momentum in the field direction (z-axis) is given by

$$L_z = m_l \frac{h}{2\pi},$$

where $m_l = 0, \pm 1, \pm 2, \ldots \ldots \pm l$. This shows that the number of possible orientations of the angular momentum in a magnetic field is $(2l + 1)$, that is, the angular momentum is space-quantised. Thus the magnetic quantum number m_l describes the space quantisation.

Electron Probability Density

As far as the exact motion of the electron within the atom is concerned, there is striking difference between old quantum theory and wave mechanics. In the old theory, the electron traces, in general, elliptic orbits whose size, eccentricity and orientation are determined by the quantisation rules. In wave mechanics, on the other hand, the electron cannot be pictured as being in a definite orbit. We can only find the relative probabilities of finding the electron at various locations. This is clearly a consequence *of* the wave nature of the electron.

The electron wave function in a hydrogen atom is given by

$$\psi_{n,l,ml} = R_{n,l}\, \Theta_{l,ml}\, \Phi_{ml},$$

where $R_{n,l,}$ $\Theta_{l,ml}$ and Φ_{ml} describe how $\psi_{n,l,ml}$ varies with r, θ and ϕ respectively. The probability density $|\psi|^2$ may therefore be written

$$|\psi|^2 = |R|^2\, |\Theta|^2\, |\Phi|^2$$

Approximate pictures of $|\psi|^2$ as functions of r, θ and ϕ for some simpler states (for which $l = 0$ and hence $m_l = (0)$ are shown in Fig.

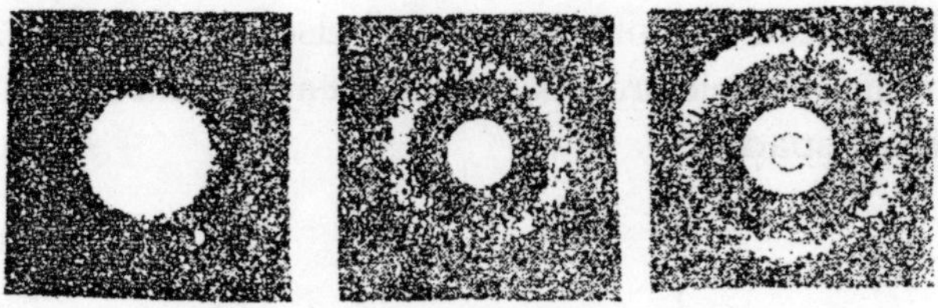

$n = 1$ $l = 0$ $n = 2$ $l = 0$ $n = 3$ $l = 0$

Since we no longer have distinct electron orbits, we speak of *electron clouds* about the nucleus. In the figure, the brightness indicates roughly the density of the electron cloud. (The density of the cloud at a specified point gives the probability of finding the electron at that point).

It is seen that the states with $l = 0$ and $n = 1, 2, 3, \ldots \ldots$ are spherically symmetrical *i.e.* the electron probability density $|\psi|^2$ has the same value at a given r in all directions. (Electron, in another states, however has angular preferences). For $n > 1$, alternate bright and dark rings appear ; the dark rings correspond to nodal surfaces where the electron is never found (zero probability)

Finally, let us see the probability of finding the electron in a hydrogen atom at a distance between r and $r + dr$ from the nucleus. This probability $P(r)$ is given by

$$P(r)\, dr = |R_{n,l}|^2\, 4\pi r^2\, dr,$$

where $4\pi r^2\, dr$ is the volume of the spherical shell whose inner radius is r and outer radius is $r + dr$. In Fig., $P(r)$ has been plotted against r for two states of hydrogen atom corresponding to $n = 1$ and $n = 2$ ($l = 0$ in both cases). It is seen that the principal maxima of the curves closely agree with the radii of the Bohr orbits ($r = a_0, 4a_0, \ldots\ldots$). Thus, whereas the electron is most likely to be found at the locations of Bohr orbits, it has a finite probability of being found elsewhere. Hence, in wave mechanics the electron is, so to speak, smeared out over the whole of space.

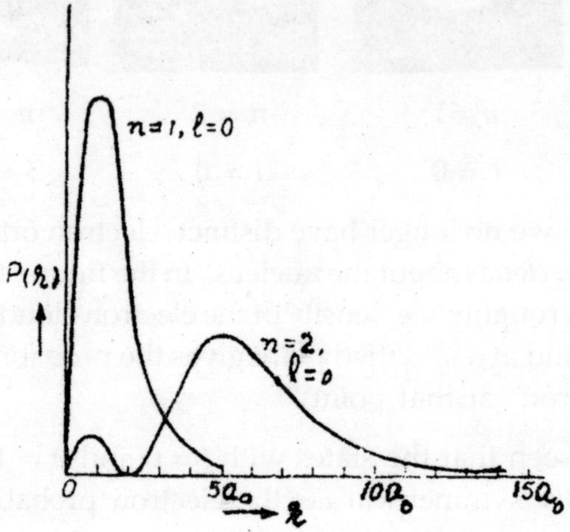

Normal State of the Hydrogen Atom

The properties of the hydrogen atom in its normal state (n — I, l = 0, ml = 0) are described by the wave function

$$\psi_{1,0,0} = R_{1,0}\,(r)\;\Theta_{0',\,0}\,(\theta)\;\Phi_0\,(\phi)$$

$$= \left(\frac{2}{a^{3/2}}\,e^{-r/a_0}\right)\left(\frac{1}{\sqrt{2}}\right)\left(\frac{1}{\sqrt{(2\pi)}}\right)$$

$$= \frac{1}{\sqrt{(\pi a_0^3)}}\,e^{-/a_0}$$

(i) The probability distribution function for the electron relative to the nucleus is given by

$$|\psi|^2 = \frac{1}{\pi a_0^3}\,e^{-2r/a_0},$$

which is independent of θ and ϕ. Hence *the normal hydrogen atom is spherically symmetrical.*

(*ii*) The probability that the electron will lie in the volume element $d\tau = r^2\, dr \sin\theta\, d\theta\, d\phi$ is given by

$$|\psi|^2\, d\tau = \frac{1}{\pi a_0^3}\left(e^{-2r/a_0}\right) r^2\, dr \sin\theta\, d\theta\, d\phi.$$

Therefore, the probability that the electron will lie between the distances r and $r + dr$ from the nucleus irrespective of its angular distribution is given by

$$P(r)\, dr = \frac{1}{\pi a_0^3}\left(e^{-2r/a_0}\right) r^2\, dr \int_0^{\pi} \sin\theta\, d\theta \int_0^{2\pi} d\phi$$

$$= \frac{1}{\pi a_0^3}\left(e^{-2r/a_0}\right) r^2\, dr\, (2)\, (2\pi)$$

$$= \frac{4}{a_0^3}\left(e^{-2r/a_0}\right) r^2\, dr.$$

A plot of radial distribution function $P(r)$ against r is shown in Fig. We see that the electron has a large probability of remaining within about 1 Å from the nucleus. This means that the size of the hydrogen atom is about the same as given by Bohr's theory. The distribution function is, however, quite different from that for a Bohr orbit of radius a_0, which would be zero everywhere except at $r = a_0$.

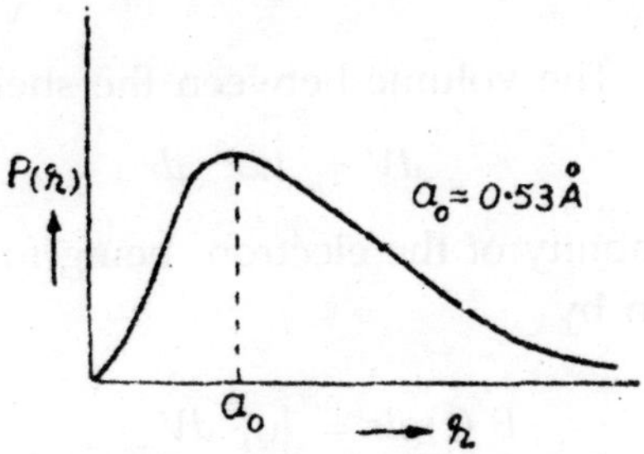

(*iii*) The most probable distance of the electron from the nucleus, *i.e.* the value of r at which $P(r)$ is maximum, is obtained by equating $\frac{dP(r)}{dr}$ to zero. That is

$$\frac{d}{dr}\left[\frac{4}{a_0^3}\left(e^{-2r/a_0}\right)r^2\right] = 0$$

or $$\frac{4}{a_0^3}\left[e^{-2r/a_0}(2r) + r^2 c^{-2r/a_0}(-2/a_0)\right] = 0$$

or $$2r + r^2(-2/a_0) = 0$$

or $$r = a_0 .$$

Thus the most probable distance of the electron from the nucleus in the normal state of hydrogen atom is equal to the Bohr's radius.

Thus, whereas in the old quantum theory the electron is allotted a definite orbit, in wave mechanics it has only a certain probability of being found at a given location. The probability is, however, maximum at points on the Bohr's orbit.

PROBLEMS

1. Consider two hypothetical spherical shells centred on the nucleus of a hydrogen atom with radii r and r + dr. What is the probability P (r) that the electron will lie between these shells, as a function of r ? The motion of the electron in the atom is described by the wave function $\psi \frac{1}{\sqrt{(\pi a_0^3)}} e^{-r/a_0} \cos \omega t$.

Solution: The volume between the shells is

$$dV = 4\pi r^2 \, dr .$$

The probability of the electron being found between the shells is given by

$$P(r)\,dr = |\psi|^2 \, dV$$

$$= \left(\frac{1}{\pi a_0^3} e^{-2r/a_0} \cos^2 \omega t\right) 4\pi r^2 \, dr.$$

The average probability is obtained by replacing $\cos^2 \omega t$ by its average value which is $\frac{1}{2}$. Thus

$$P(r) = \left\{\frac{1}{\pi a_0^3} e^{-2r/a_0} \left(\frac{1}{2}\right)\right\} 4\pi r^2$$

$$= \frac{2r^2}{a_0^3} e^{-2r/a_0}$$

2. *Show that the degree of degeneracy of the nth energy level is the hydrogen atom in n*^{2}.

Solution : The energy values given by

$$E = -\frac{\mu e^4}{8 \epsilon_0^2 h^2}\left(\frac{1}{n^2}\right)$$

depend only on n and hence are degenerate with respect to both l and ml (*i.e.* the states having same n but different l's or m_l's have same energy).

For each value of n, l can take n values from 0 to $n - 1$, and for each value of l there are $(2l + 1)$ values of m_l ($= 0, \pm 1, \pm 2, \ldots\ldots \pm l$) Therefore, the degeneracy of the nth energy level is

$$\sum_{i=0}^{n-1}(2l+1)$$

$$= 1 + 3 + 5 + \ldots\ldots + (2n - 1);$$

a total of n terms

$$= \frac{n}{2}\,[2(1) + (n - 1)\,2] = n^2.$$

5

Development of Wave Machines

Situation Leading to Wave Mechanics

Limitations of the Classical Mechanics: Bohr model of the hydrogen atom accounts for the hydrogen line spectrum, but with a number of shortcomings. For example, it does not explain the fine structure of spectral lines and the intensity variation in them. Moreover, its use is limited to the (simplest) hydrogen and hydrogen-like atoms ; it cannot explain the spectra of complex atoms. Finally, but most important, it does not tell how electrons interact with atoms, and atoms with atoms, to give the observed properties of matter.

The cause of these shortcomings goes back to the limitations of the classical (Newtonian) mechanics itself which is usually applied to explain the properties of systems. The classical mechanics is based on the assumption that in any system of particles it is possible to measure precisely the quantities like position, velocity, momentum, acceleration, etc., which characterise the state of the system at any instant. The equations of motion then enable us to determine these quantities, and hence the state of the system, at all later instants. The assumption is valid in our every-day experience. This is why classical mechanics provides correct explanation for the behaviour of macroscopic systems.

The assumption is, however, in conflict with the uncertainty principle which becomes important in the atomic realm. According to this principle, the position and momentum of a particle cannot be accurately measured at the same time. The measurement of one quantity introduces an uncertainty into the other. Therefore, classical mechanics which assumes both to have definite values at all instants is not valid for atomic systems.

This is the reason that the Bohr model, which was just a classical atomic model in which quantization was introduced arbitrarily, failed to give a complete explanation of the behaviour of atomic systems. In fact, the success of classical mechanics in case of macroscopic systems is due to the fact that such systems consist of so many individual atoms that departures from average behaviour are unnoticeable. But truely speaking, classical mechanics must be replaced by a theory of greater generality. Wave mechanics, developed by Schrodinger, is a correct replacement. Classical mechanics is, in fact, an approximation of wave mechanics.

Wave mechanics is consistent with the uncertainty principle. Classically we say that the radius of the electron's orbit in a ground-state hydrogen atom is always 053 A. Wave mechanics, however, states that 0.53 Å is the *most probable* radius. If we perform an experiment, most trials will yield a different value, either larger or smaller, but the value most likely to be found will be 0.53 Å.

Operators: We know that a function is a rule by means of which, from a given number we can find another number. Similarly, *an operator is a rule by means of which, from a given function we can find another function.*

For example, an operator may be the multiplication of a function $f(x)$ by the independent variable x, that is,

$$\alpha(x) = xf(x).$$

Similarly, an operator β may be the differentiation of a function $f(x)$ with respect to the independent variable x, that is,

$$\beta f(x) = \frac{d}{dx} f(x) = f'(x)$$

We can see that

$$\alpha\beta[f(x)] = x\frac{d}{dx} f(x) = xf'(x)$$

and $$\beta\alpha[f(x)] = \frac{d}{dx}\left[xf(x)\right] = xf'(x) + f(x).$$

Thus $$\alpha\beta \neq \beta\alpha,$$

If, however, the operators α and β are such that

$$\alpha\beta = \beta\alpha,$$

then α and β are said to 'commute'.

Eigenfunctions and Eigenvalues

If there is a class of functions which, when operated by an operator α, are merely multiplied by some constant (say) a, that is, if

$$\alpha\, f(x) = af(x).$$

then these functions are called 'eigenfunctions' of the operator α, and the various possible values of a are called 'eigenvalues' of the operator.

For example, let us consider the function $\sin 2x$. If it is operated by an operator $-\frac{d^2}{dx^2}$, the result is

$$-\frac{d^2}{dx^2}(\sin 2x) = 4\ (\sin 2x).$$

Thus sin $2x$ is the eigenfunction and 4 is the eigenvalue of the operator $-\frac{d^2}{dx^2}$. In wave mechanics, the allowed eigenfunctions are continuous, finite and single-valued.

If two operators commute then there exists a set of functions which are simultaneously eigenfunctions of both the operators.

Properties of Functions and Operators: The following properties of functions and operators are important in wave mechanics :

(1) *An operator* α *operating on any two functions* ψ_1 *and* ψ_2 *is 'Hermitian', if*

$$\int \psi^*_2 (\alpha\psi_1)\, d\tau = \int \psi_1 (\alpha^* \psi_2{}^*)\, d\tau$$

The Hermitian operators have the property that their eigenvalues are always real.

Proof : Let ψ be an eigenfunction of an operator α with eigenvalue a. Then

$$\alpha\,\psi = a\,\psi\,.$$

Taking complex conjugate of every quantity, we have

$$\alpha^*\ \psi^* = a^*\ \psi^*$$

Now $$\int \psi^* (\alpha\psi)\, d\tau = \int \psi^* (a\psi)\, d\tau = a^* \int \psi\psi^*\, d\tau$$

and $$\int \psi (\alpha^* \psi^*)\, d\tau = \int \psi (a^* \psi^*)\, d\tau = a^* \int \psi\,\psi^*\, d\tau$$

If α is Hermitian, then

$$\int \psi^* (\alpha\psi) d\tau = \int \psi (\alpha^* \psi^*) d\tau$$

$\therefore$ $$a \int \psi^* \psi\, d\tau = a^* \int \psi\psi^*\, d\tau$$

or $$a = a^*,$$

which is true only when a is real.

(2) *An operator* α *is said to be linear if*

$$\alpha(\psi_1 + \psi_2) = \alpha\psi_1 + \alpha\psi_2.$$

(3) *Two wave functions* ψ_1 *and* ψ_2 *are 'orthogonal' if*

$$\int \psi_2^* \psi_1 \, d\tau = 0$$

An important theorem regarding orthogonality is that two eigenfunctions of a Hermitian operator are orthogonal to each other if the corresponding eigenvalues are unequal.

Proof: Let ψ_1 and ψ_2 be two eigenfunctions of an operator α with eigenvalues a_1 and a_2. Then

$$\alpha\psi_1 = a_1\psi_1$$

and $$\alpha\psi_2 = a_2\psi_2$$

We can write

$$\int \psi_2^* \alpha\psi_1 \, d\tau = a_1 \int \psi_2^* \psi_1 \, d\tau$$

Since α is Hermitian, we have

$$\int \psi_2^* (\alpha\psi_1) \, d\tau = \int \psi_1 (\alpha^* \psi_2^*) \, d\tau$$

$$= a_2^* \int \psi_1 \psi_2^* \, d\tau$$

$$[\because \alpha^* \psi_2^* = a_2^* \psi_2^*]$$

$$= a_2 \int \psi_1 \psi_2^* \, d\tau$$

because $a_2^* = a_2$ (as α is Hermitian). Therefore

$$a_1 \int \psi_2^* \psi_1 \, d\tau = a_2 \int \psi_1 \psi_2^* \, d\tau$$

or $$(a_1 - a_2) \int \psi_2^* \psi_1 \, d\tau = 0$$

or $$\int \psi_2^* \psi_1 \, d\tau = 0, \text{ provided } a_1 \neq a_2$$

which is the condition of orthogonality.

Postulates of Wave Mechanics

Let us consider a system consisting of n particles moving in a conservative field of force: Classically, the system can be described at any time t in terms of the positional coordinates $q_1, q_2, \ldots\ldots q_n$ + and the linear momentum coordinates $p_1, p_2, \ldots\ldots\ldots p_n$ of the particles. The wave mechanical description of the system is based on certain postulates which are as follows :

I. Postulate: There is an associated wave function ψ (q_1, q_2, ...q_n, t) with the system, and any state of the system is described as fully as possible by this wave function. This is the 'state function of the system. *It has the property that* $\psi\psi^* \, d\tau$ *is the probability for the system having q values for lying within the volume element* $d\tau$ *at time t (i.e.* q_1 has a value between q_1 and $q_1 + dq_1$, etc. at time t). Since each q must have some value, the total probability must be unity, that is,

$$\int \psi\psi^* \, d\tau = 1.$$

II. Postulate: With every dynamical variable there is associated an operator. If the variable is a physically measurable quantity (such as the position coordinate of the particle, the energy of the system, etc.) then the associated operator is Hermitian.

The correct operators to be associated with given measurable variables are as follows :

(*i*) If the variable is a positional coordinate q_i or time t, then the operator is multiplication by itself.

(*ii*) If the variable is a linear momentum coordinate p_i, then the operator is $-\dfrac{ih}{2\pi} \dfrac{\partial}{\partial q_i}$, where q_i is conjugate to p_i.

(*iii*) If the variable, say A, is expressible in terms of q's, p's and t, the operator is found by substituting the operators for q's, p's and t in the algebraic equation for A and arranging them so that the resulting operator is Hermitian.

III. Postulate: The wave functions $\psi\ (q,\ t)$ satisfy the equation

$$H\left(q, -\frac{ih}{2\pi}\frac{\partial}{\partial q}, t\right)\psi(q,t) = \frac{ih}{2\pi}\frac{\partial}{\partial t}\psi(q,t),$$

where H is the Hamiltonian operator (or energy operator) for the system. The above equation is known as Schrodinger's time-dependent equation.

IV. Postulate: The precise value a of a dynamical variable A for a system is given by

$$\alpha\psi\ (q,\ t) = a\psi\ (q,\ t)$$

where α is the corresponding operator and $\psi\ (q,\ t)$ is the eigenfunction of α, a is called the 'expectation value'.

Particularly, the energy value E of the system is given by

$$H\left(q, -\frac{ih}{2\pi}\frac{\partial}{\partial q}, t\right)\psi(q,t) = E\psi\ (q,\ t),$$

where H is the Hamiltonian operator.

If this equation is to be consistent with Schrodinger's time-dependent equation (III postulate), then $\psi\ (q,\ t)$ must be of the form

$$\psi(q,\ t) = \psi(q)\ e^{-2\pi i\ Et/h}$$

where $\psi\ (q)$ is a solution of the equation

$$H\left(q, -\frac{ih}{2\pi}\frac{\partial}{\partial q}, t\right)\psi(q) = E\psi(q)$$

This is Schrodinger's equation for a steady state.

As an example illustrating the use of these postulates, let us consider a system consisting of a particle of mass m moving in a potential field $V\ (x,\ y,\ z)$. The kinetic energy is

$$K = \frac{1}{2m}\left(p_x^2 + p_y^2 + p_z^2\right).$$

Therefore, the classical total energy (Hamiltonian) function is

$$H = \frac{1}{2m}\left(p_x^2 + p_y^2 + p_z^2\right) + V(x, y, z).$$

From the above postulates, the Hamiltonian operator is

$$H\left(q, -\frac{ih}{2\pi}\frac{\partial}{\partial q}, t\right)$$

$$= \frac{1}{2m}\left[\left(-\frac{ih}{2\pi}\frac{\partial}{\partial x}\right)^2 + \left(-\frac{ih}{2\pi}\frac{\partial}{\partial y}\right)^2 + \left(-\frac{ih}{2\pi}\frac{\partial}{\partial z}\right)^2\right] + V(x, y, z)$$

$$= -\frac{h^2}{8\pi^2 m}\nabla^2 + V(x, y, z)$$

Substituting it in Schrodinger's time-dependent equation (postulate III), we get

$$\left[-\frac{h^2}{8\pi^2 m}\nabla^2 + V(x, y, z)\right]\psi(x, y, z, t) = \frac{ih}{2\pi}\frac{\partial}{\partial t}\psi(x, y, z, t)$$

Let a solution of this equation be

$$\psi(x, y, z, t) = \psi(x, y, z)\, e^{-2\pi i Et/h}$$

Then we obtain

$$\left[-\frac{h^2}{8\pi^2 m}\nabla^2 + V(x, y, z)\right]\psi(x, y, z)\, e^{-2\pi i Et/h}$$

$$= \frac{ih}{2\pi}\frac{\partial}{\partial t}\psi(x, y, z)\, e^{-2\pi i Et/h}$$

$$= \frac{ih}{2\pi}\psi(x, y, z)\, e^{-2\pi i Et/h}\left(-2\pi i\, E/h\right)$$

$$= E\psi(x, y, z)\, e^{-2\pi i Et/h}$$

or $$\left[-\frac{h^2}{8\pi^2 m}\nabla^2 + V(x,y,z)\right]\psi(x,y,z)$$

$$= E\psi\ (x, y, z).$$

This is Schrodinger's steady-state equation. Thus if $\psi(x,y,z)$ is a solution of this equation, then it is an eigenfunction of the Hamiltonian (energy) operator and describes a state with the precise value E for the energy.

The last equation can be written as

$$\nabla^2\psi + \frac{8\pi^2 m}{h^2}(E - V)\psi = 0.$$

Schrodinger's Time-dependent Wave Equation: In 1926, Schrodinger presented his famous wave equation as a development of de Broglie ideas of the wave properties of matter. The Schrodinger's equation is the fundamental equation of wave mechanics in the same sense as the Newton's second law of motion of classical mechanics. It is the differential equation for the de Broglie waves associated with particles and describes the motion of particles.

Schrodinger introduced a mathematical function ψ which is the variable quantity associated with the moving particle, and is a complex function of the space coordinates of the particle and the time, ψ is called the "wave function" as it characterises the (de Broglie) waves associated with the particle. It is postulated that ψ has the form of a solution of the classical wave equation.

The differential equation representing a one-dimensional wave motion is

$$\frac{\partial^2 y}{\partial x^2} = \frac{1}{v_p^2}\frac{\partial^2 y}{\partial t^2},$$

where v, is the phase velocity of waves moving along the x-axis. This equation holds for waves in *any* medium in which the wave velocity v_p is independent of the character of the waves and depends only upon the properties of the medium. The general solution of this equation is

$$y = Ae^{-i\omega\{t-(x/v_p)\}}$$

where A is the constant amplitude and ω is the constant angular frequency. The wave equivalent of a "free" particle moving in the $+x$ direction may therefore be described by a corresponding equation

$$\psi\,(x,\,t) = Ae^{-i\omega\{t-(x/v_p)\}}$$

We know that $\omega = 2\,\pi\,v$ and $v_p = v\lambda$, where v and A are frequency and wavelength respectively. Thus we can write

$$\psi(x,\,t) = Ae^{-2\pi i\,\{vt-(x/\lambda)\}}\,.$$

Since $v = \dfrac{E}{h}$ (Planck's relation)

and $\lambda = \dfrac{h}{p}$ (de Broglie relation),

we have $\quad \psi(x,\,t) = Ae^{-(2\pi i/h)(Et-px)}$ ... (*i*)

This is the wave description of a "free" particle of total energy E and momentum p moving in the $+\,x$ direction. From this we can easily obtain the differential equation for ψ, which can be solved for particles subjected to various restrictions.

Differentiating eq. (*i*) twice with respect to x, we get

$$\frac{\partial^2\psi}{\partial x^2} = Ae^{-(2\pi i/h)(Et-px)}\left(\frac{2\pi i}{h}p\right)^2$$

$$= -\frac{4\pi^2p^2}{h^2}\psi\,.$$

Again, differentiating eq. (*i*) once with respect to t, we get

$$\frac{\partial \psi}{\partial t} = Ae^{-(2\pi i/h)(Et - px)}\left(-\frac{2\pi i}{h}E\right)$$

$$= -\frac{2\pi i E}{h^2}\psi .$$

These expressions give

$$p^2 = -\frac{1}{\psi}\frac{h^2}{4\pi^2}\frac{\partial^2 \psi}{\partial x^2} \qquad \text{... } (ii)$$

and

$$E = -\frac{1}{\psi}\frac{h}{2\pi i}\frac{\partial \psi}{\partial t} = \frac{1}{\phi}\frac{ih}{2\pi}\frac{\partial \phi}{\partial t} \qquad \text{... } (iii)$$

In non-relativistic mechanics, the total energy E of a particle is the sum of its kinetic energy $p^2/2m$ and potential energy V. That is

$$E = \frac{p^2}{2m} + V$$

Substituting in this eq. for p^2 and E from (*ii*) and (*iii*), we get

$$\frac{1}{\psi}\frac{ih}{2\pi}\frac{\partial \psi}{\partial t} = -\frac{1}{\psi}\frac{h^2}{8\pi^2 m}\frac{\partial^2 \psi}{\partial x^2} + V$$

or

$$\boxed{-\frac{h^2}{8\pi^2 m}\frac{\partial^2 \psi}{\partial x^2} + V\psi = \frac{ih}{2\pi}\frac{\partial \psi}{\partial t}}$$

This is the "time-dependent" form of Schrodinger's equation. Any restriction on particle's motion will affect the potential energy V which, in general, is a function of x and t. Once V is known, the equation may be solved for ψ. This equation is used in explaining non-stationary phenomenon e.g. an electronic transition between two states of an atom.

Schrodinger's Time-independent Wave Equation: In many cases the potential energy of a particle does not depend upon time and is a function of the position only. When this is the case, the Schrodinger's equation may be simplified.

Let us again start from eq. (*i*) and write it as

$$\psi(x, t) = \left(Ae^{2\pi ipx/h}\right)\left(e^{-2\pi iEt/h}\right)$$

in which the variables x and t have been separated. If the spatial part is

$$f(x) = Ae^{2\pi ipx/h},$$

then the above equation can be written as

$$\psi(x, t) = \psi(x)e^{-2\pi iEt/h},$$

Now, by differentiation twice with respect to x, we get

$$\frac{\partial^2\psi}{\partial x^2} = \frac{d^2\psi}{dx^2}e^{-2\pi iEt/h}$$

and differentiation with respect to time gives

$$\frac{\partial\psi}{\partial t} = \psi e^{-2\pi iEt/h}\left(-\frac{2\pi iE}{h}\right)$$

Substituting these values of ψ, $\partial^2\psi/\partial x^2$ and $\partial\psi/\partial t$ into the time-dependent form of Schrodinger equation (*iv*), we get

$$-\frac{h^2}{8\pi^2 m}\frac{d^2\psi}{dx^2}e^{-2\pi iEt/h} + Va_2\int\psi_1\psi_2{}^*d\psi e^{-2\pi Et/h}$$

$$= \frac{ih}{2\pi}\left(-\frac{2\pi iE}{h}\right)\psi e^{-2\pi iEt/h}$$

or $-\dfrac{h^2}{8\pi^2 m}\dfrac{d^2\psi}{dx^2} + V\psi = E\psi$

or $\boxed{\dfrac{d^2\psi}{dx^2} + \dfrac{8\pi^2 m}{h^2}(E - V)\psi = 0}$

This is the "steady state" or "time-independent" form of Schrodinger equation.

Schrodinger's Time-independent Equation: Schrodinger's equation is a differential equation for the de Broglie waves associated with a particle and thus describes the motion of the particle. The variable quantity associated with the moving particle is a mathematical function ψ, called the wave function.

For steady-state systems, ψ is a function of spatial coordinates only.

A one-dimensional equation for the steady waves associated with a particle may be written as

$$\psi = A \sin \frac{2\pi x}{\lambda}, \qquad \dots (i)$$

where the function ψ represents the amplitude of the wave for any given value of x, A is the maximum amplitude and λ is the wavelength. Differentiating ψ twice with respect to x, we get

$$\frac{d^2\psi}{dx^2} = -\frac{4\pi^2}{\lambda^2} A \sin \frac{2\pi x}{\lambda} = -\frac{4\pi^2}{\lambda^2}\psi \qquad \dots (ii)$$

Now, the wavelength associated with a particle of mass m moving with velocity v is given by

$$\lambda = \frac{h}{mv}$$

so that

$$\frac{1}{\lambda^2} = \frac{m^2v^2}{h^2} = \frac{2m\left(\frac{1}{2}mv^2\right)}{h^2} \qquad \dots (iii)$$

For non-relativistic motion of the particle, $\frac{1}{2}mv^2$ is the particle's kinetic energy K. Thus if E be the total energy of the particle and V the potential energy, then we can write

$$\frac{1}{2}mv^2 = K = E - V.$$

Making this substitution in eq. (*iii*), we get

$$\frac{1}{\lambda^2} = \frac{2m}{h^2}(E - V).$$

Now, substituting this value of $\frac{1}{\lambda^2}$ in eq. (*ii*), we obtain

$$\frac{d^2\psi}{dx^2} = -\frac{8\pi^2 m}{h^2}(E - V)\psi$$

or $$\frac{d^2\phi}{dx^2}+\frac{8\pi^2 m}{h^2}(E-V)\,\phi=0$$

This is the time-independent (steady state) Schrodinger equation for a particle of mass m, total energy E, potential energy V, moving along the x-axis.

For the same particle moving in three-dimensional space, the equation becomes

$$\frac{\partial^2\psi}{\partial x^2}+\frac{\partial^2\psi}{\partial y^2}+\frac{\partial^2\psi}{\partial z^2}+\frac{8\pi^2 m}{h^2}(E-V)\,\psi = 0$$

or $$\nabla^2\psi+\frac{8\pi^2 m}{h^2}(E-V)\,\psi=0$$

For a *free* particle $V = 0$. Therefore, the Schrodinger's equations for a free particle is

$$\nabla^2\psi+\frac{8\pi^2 mE}{h^2}\psi = 0$$

Physical Interpretation of Wave Function ψ: A satisfactory interpretation of the wave function ψ associated with a moving particle was given by Born in 1926. He postulated that the square of the magnitude of the wave function, $|\psi|^2$ (or $\psi\psi^*$ if ψ is complex) evaluated at a particular point represents the probability of finding the particle at that point, $|\psi|^2$ is called the probability density and ψ is the 'probability amplitude'.

According to this interpretation, the probability of finding the particle within an element of volume $d\tau$ is $|\psi|^2 d\tau$. Since the particle is certainly somewhere, the integral of $|\psi|^2 d\tau$ over the whole space must be unity, that is

$$\int_{-\infty}^{+\infty}|\psi|^2\,d\tau = 1.$$

A wave function that obeys this equation is said to be 'normalised', Every acceptable wave function must be normalizable.

Besides being normalizable, an acceptable wave function ψ must fulfil the following requirements :

(i) *It must be Finite Everywhere*: If, for instance, ψ is infinite for a particular point, it would mean an infinitely large probability of finding the particle at that point. This would violate the uncertainty principle. Therefore ψ must have a finite or zero value at any point.

(ii) *It must be Single-valued:* If ψ has more than one, value at any point, it would mean more than one value of probability of finding the particle at that point which is obviously rediculous.

(iii) *It must be Continuous and have a Continuous First Derivative Everywhere:* This is necessary from the Schrodinger equation itself which shows that $d^2\psi/dx^2$ must be finite everywhere. This can be so only if $d\psi/dx$ has no discontinuity at any boundary where potential changes. Furthermore, the existence of $d\psi/dx$ as a continuous function implies that ψ too is continuous across a boundary.

These requirements which must be fulfilled by an acceptable wave function carry great significance. When the Schrodinger steady-state equation for a given system is solved to obtain a wave function which fulfils these requirements then we find that the equation can be salved only for certain values of the energy of the system. Thus energy quantization appears in wave mechanics as a natural feature of the solution of the wave equation. The values of energy for which Schrodinger equation can be solved are called "eigenvalues" and the corresponding (acceptable) wave functions are called "eigenfunctions".

The Schrodinger equation for a particle enclosed its a one-dimensional rigid box of side L and obtain its eigenvalues may be solved. A graph of its first three eigenfunctions is drawn. The probability of finding the particle at different points in the box when it is in different States is discussed.

Particle in a Rigid One-dimensional Box (Infinite Potential Well): Let us consider a particle restricted to move along the x-axis between $x = 0$ and $x = L$ by ideally reflecting, infinitely high walls of a box (Fig.). Suppose that the potential energy V of the particle is zero inside the box, but rises to infinity on the outside, that is,

$$V = 0 \quad \text{for } 0 \leq x \leq L$$

and

$$V = \infty \text{ for } x < 0 \text{ and } x > L.$$

In such a case, the particle is said to be moving in an infinitely deep potential well.

The Schrodinger equation for the particle within the box ($V = 0$) is

$$\frac{d^2\psi}{dx^2} + \frac{8\pi^2 mE}{h^2}\psi = 0 \qquad \text{... } (i)$$

If we put

$$\frac{8\pi^2 mE}{h^2} = k^2, \qquad \text{... } (ii)$$

then we have

$$\frac{d^2\psi}{dx^2} + k^2\psi = 0 \qquad \text{... } (iii)$$

The general solution of this differential equation is of the form

$$\psi(x) = A \sin kx + B \cos kx, \qquad \text{... } (iv)$$

where the constants A and B are to be determined by the boundary conditions.

Since the particle cannot have infinite energy, it cannot exist outside the box. Therefore, the wave function ψ must be zero outside the box. ψ must be also zero at the walls, that is, at $x = 0$ and $x = L$, for otherwise there would be discontinuities at the walls.

The boundary condition $\psi(x) = 0$ at $x = 0$ applied to eq. (*iv*) requires

$$B = 0,$$

and the condition $\psi\ (x) = 0$ at $x = L$ requires

$$A \sin kL = 0$$

or $$kL = n\pi$$

or $$k = \frac{n\pi}{L},$$

where $n = 1, 2, 3, ...$ ($n = 0$ is not admissible because it yields ψ zero everywhere which means that the particle is nowhere). Substitution of this value of k in eq. (*ii*) shows that the energy E can have only the values

$$\boxed{E_n = \frac{n^2h^2}{8mL^2}.} \quad n = 1, 2, 3, ... \qquad ...(v)$$

Thus we see that in a potential well the particle cannot have an arbitrary energy, but can take only certain discrete energy values corresponding to $n = 1, 2, 3......$ These are the 'eigenvalues' of the particle in the well and constitutes the energy levels of the system (Fig. below). The integer n corresponding to the energy level E_n is called its 'quantum number'.

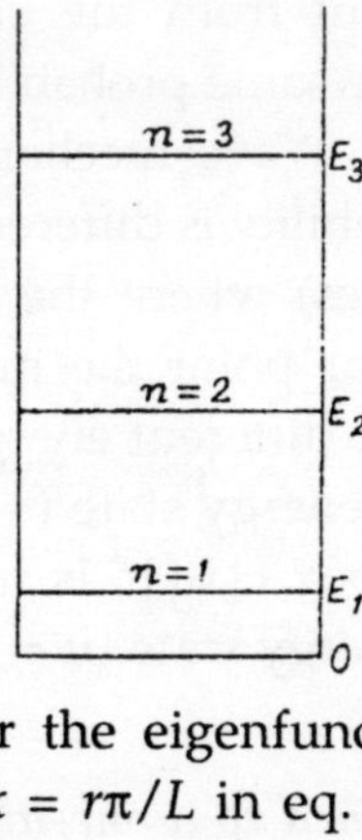

Let us now consider the eigenfunctions of the particle. Substituting $B = 0$ and $k = r\pi/L$ in eq. (*iv*), we find that the allowed solutions of the Schrodinger equation are

$$\psi_n(x) = A \sin \frac{n\pi x}{L}.$$

To find the value of A, we use the normalization condition

$$\int_{-\infty}^{+\infty} |\psi_n(x)|^2 \, dx = 1$$

In the present case, this will be written as

$$A^2 \int_0^L \sin^2 \frac{n\pi x}{L} dx = 1$$

This gives

$$A^2 \left(\frac{L}{2}\right) = 1 \text{ or } A = \sqrt{\left(\frac{2}{L}\right)}.$$

The normalised eigenfunctions of the particle are therefore

$$\boxed{\psi_n(x) = \sqrt{\left(\frac{2}{L}\right)} \sin \frac{n\pi x}{L}.}$$

The first three eigenfunctions ψ_1, ψ_2, ψ_3 together with the probability densities $|\psi_1|^2$, $|\psi_2|^2$ $|\psi_3|^2$ are shown in same figures respectively. It is obvious that the wave mechanical result is very different from the classical result. Classical mechanics predicts the same probability for the particle being anywhere in the box. Wave mechanics, on the other hand, predicts that the probability is different at different points and there are points (nodes) where the particle is never found. Further, at a particular point the probability of finding the particle is different for different energy states. For example, a particle in the lowest energy state ($n = 1$) is most likely to be in the middle of the box ($|\psi_1|^2$ is maximum there); while a particle in the next energy state ($n = 2$) is never there ($(|\psi_2|)^2$ is zero there).

A particle is confined to a one-dimensional box of side L. Expressions for its eigenfunctions and eigenvalues are written. That the minimum energy of this particle cannot be zero is shown. The finite minimum energy on the basis of uncertainty principle is justified.

Eigenfunctions and Eigenvalues of a Particle in a Box: The eigenfunctions and eigenvalues of a particle of mass m moving along the x-axis in a rigid box of side L are given by

$$\psi_n(x) = A \sin \frac{n\pi x}{L},$$

and $$E_n = \frac{n^2 h^2}{8mL^2}, \quad \text{where } n \text{ is an integer.}$$

It is obvious that the quantum number n cannot be zero because this would make $\psi(x) = 0$ everywhere in the box which means that the particle is not present at all. Now, the exclusion of $n = 0$ means that *the particle cannot have zero energy.* This quantum mechanical result has no counterpart in classical mechanics where all energies including zero are possible.

The lowest value of n is 1, and thus the lowest allowed value of the energy is

$$E_1 = \frac{h^2}{8mL^2}.$$

This is in consistent with the uncertainty principle. Since the particle is confined to the box, the uncertainty in its position is equal to the length of the box, that is,

$$\Delta x = L.$$

The uncertainty in its momentum must therefore be, by relation $\Delta x \; \Delta p \geq h/4\pi$,

$$\Delta p \geq \frac{h}{4\pi L},$$

and the uncertainty in the (kinetic) energy would be

$$\Delta E = \frac{(\Delta p)^2}{2m} \geq \frac{h^2}{32\pi^2 mL^2}$$

The minimum energy must be of the same order as the uncertainty in energy. This is consistent with the value obtained by putting $n = 1$ in the energy expression. Thus the uncertainty principle provides confirmation that $E = 0$ is not admissible.

In our daily life we come across particles of large mass and systems of large dimensions. In such case the allowed energy states do not appear to be quantized and for all practical purposes, form a continuum. To show this, let us find the spacings between two adjacent energy levels for different values of the mass of a particle moving in boxes of different sizes.

The difference in energies of two adjacent levels is

$$\Delta E_n = E_{n+1} - E_n$$

$$= \frac{(n+1)^2 h^2}{8m L^2} - \frac{n^2 h^2}{8mL^2}$$

$$= \frac{h^2}{8mL^2}(2n+1)$$

Let us take m to be the mass of a molecule ($\approx 10^{-26}$ kg) and L to be 10 cm (= 0.1 meter). That is, for the molecule of a gas in a vessel, we have

$$\Delta E_n = \frac{\left(6.62 \times 10^{-34}\right)^2 (2n+1)}{8 \times 10^{-26} \times (0.1)^2}$$

$$= 5.5 \times 10^{-40} \ (2n + 1) \text{ joule}$$

$$= \frac{5.5 \times 10^{-40} \ (2n+1)}{1.6 \times 10^{-19}}$$

$$= 3.4 \times 10^{-21} \ (2n + 1) \text{ eV}$$

The permissible energy levels are so very close together that they may be regarded as being practically continuous. Although energy quantization occurs in principle, it will not affect the motion of the molecules. Hence in the domain of everyday experience quantum effects are imperceptible. This accounts for the success of Newtonian mechanics in every-day life.

A similar result is obtained if we take m to be the mass of an electron (9×10^{-31} kg) but L same as above. In this case of free electrons in a metal, we have

$$\Delta E_n = 3.8 \times 10^{-17} (2n + 1) \text{eV}.$$

Thus in a trap of macroscopic size the energy of an electron behaves in classical manner; it can take continuous values.

An absolutely different result is, however, obtained for an electron moving in a region of atomic dimensions ($m = 9 \times 10^{-31}$ kg, $L = 1$ Å $= 10^{-10}$ meter). In this case

$$\Delta E_n = \frac{\left(6.62 \times 10^{-34}\right)^2 (2n+1)}{8 \times \left(9 \times 10^{-31}\right) \times \left(10^{-10}\right)^2}$$

$$= 6 \times 10^{-18} (2n + 1) \text{ joule}$$

$$= \frac{6 \times 10^{-18} (2n+1)}{1.6 \times 10^{-19}}$$

$$= 38 (2n + 1) \text{ eV}.$$

The spacing between the lowest ($n = 1$) and the next level is 114 eV, and increases in higher adjacent levels. Thus these energy levels are sufficiently far apart to make the quantization of energy conspicuous.

Thus, in passing from atomic to much-larger dimensions the results of wave mechanics approach those of classical physics.

Particle in a Three-dimensional Rigid Box: In three-dimensions, the steady-state form of Schrodinger's equation for a particle of mass m is

$$\frac{\partial^2 \psi}{\partial x^2} + \frac{\partial^2 \psi}{\partial y^2} + \frac{\partial^2 \psi}{\partial z^2} + \frac{8\pi^2 m}{h^2}(E - V)\psi = 0. \qquad \text{... } (i)$$

Suppose the particle is confined to a cubical box of side L with infinitely hard walls which are parallel to the coordinate axes. The potential energy V of the particle is zero inside the box but rises to infinity outside it. Inside the box the Schrodinger's equation is

$$\frac{\partial^2\psi}{\partial x^2}+\frac{\partial^2\psi}{\partial y^2}+\frac{\partial^2\psi}{\partial z^2}+\frac{8\pi^2 mE}{h^2}\psi = 0 \qquad \text{... } (ii)$$

To solve this equation, we must separate it into three independent equations, each of which has only one coordinate. For this, let us write the wave function $\psi(x, y, z)$ as the product of three different functions : $\psi_x(x)$, $\psi_y(y)$ and $\psi_z(z)$ which depend respectively upon x alone, y alone and z alone. That is

$$\psi(x, y, z) = \psi_x(x)\,\psi_y(y)\,\psi_z(z).$$

Taking partial derivatives, we get

$$\frac{\partial^2\psi}{\partial x^2} = \psi_y\psi_z\frac{d^2\psi_x}{dx^2},$$

$$\frac{\partial^2\psi}{\partial y^2} = \psi_x\psi_z\frac{d^2\psi_y}{dy^2}$$

and
$$\frac{\partial^2\psi}{\partial z^2} = \psi_x\psi_y\frac{d^2\psi_z}{dz^2}$$

Making the above four substitutions in eq. (*ii*), we get

$$\psi_y\psi_z\frac{d^2\psi_x}{dx^2}+\psi_x\psi_z\frac{d^2\psi_y}{dy^2}+\psi_x\psi_y\frac{d^2\psi_z}{dz^2}+\frac{8\pi^2 mE}{h^2}\psi_x\psi_y\psi_z = 0$$

Dividing throughout by $\psi_x\,\psi_y\psi_z$, we have

$$\frac{1}{\psi_x}\frac{d^2\psi_x}{dx^2}+\frac{1}{\psi_y}\frac{d^2\psi_y}{dy^2}+\frac{1}{\psi_z}\frac{d^2\psi_z}{dz^2} = -\frac{8\pi^2 mE}{h^2}$$

In this equation each term on the left-hand side is a function of a different variable, while the right-hand side is a constant. Therefore, each term on the left-hand side must be equal to a separate constant. Thus, we may write

$$\frac{1}{\psi_x}\frac{d^2\psi_x}{dx^2} = -k_x^2$$

$$\frac{1}{\psi_y}\frac{d^2\psi_y}{dy^2} = -k_y^2$$

$$\frac{1}{\psi_z}\frac{d^2\psi_z}{dz^2} = -k_z^2$$

with the condition that

$$k_x^2 + k_y^2 + k_z^2 = \frac{8\pi^2 mE}{h^2}$$

Let us consider the first of the above equations, which can be written as

$$\frac{d^2\psi_x}{dx^2} + k_x^2\psi_n = 0$$

Using the boundary conditions $\psi_n = 0$ at $x = 0$ and $x = L$, the eigenfunctions and eigenvalues corresponding to this equation will be given by (as obtained in the one-dimensional case)

$$\psi_n = A_x \sin\frac{n_x\pi x}{L}$$

and
$$E_x = \frac{n_x^2 h^2}{8mL^2}$$

where $n_x = 1, 2, 3, ...$ Similar expressions will hold for ψ_y, E_y, ψ_x and E_x.

We therefore have for the complete wave function

$$\psi = \psi_x\psi_y\psi_z$$

$$= A_x A_y A_z \sin\frac{n_x\pi x}{L}\sin\frac{n_y\pi y}{L}\sin\frac{n_z\pi z}{L},$$

where $A_xA_yA_z$ is a normalization constant whose value must be such that $\int|\psi|^2 dE = 1$.

The possible energies of the particle are

$$E = E_x + E_y + E_z$$

$$= \left(n_x^2 + n_y^2 + n_z^2\right)\frac{h^2}{8mL^2},$$

where $n_z = 1, 2, 3.........$, $n_v = 1, 2, 3......$, $n_z = 1, 2, 3..........$

Penetration of Particles through Rectangular Potential Barrier (Quantum Tunnelling): The behaviour of a particle approaching a region in which its potential energy changes suddenly is a typical wave mechanical problem. The situation can be realised physically, for example, with an electron moving towards a region of reverse electric field. Such a field is best described in terms of its potential and is called 'potential-barrier'.

Suppose a particle moving along the x-axis from left to right meets a rectangular potential barrier of height V_0 and of width L which is bounded by sharp potential jumps : one from zero to a finite value V_0 , and the other from V_0 back to zero (Fig. below). We have three different regions separated by discontinuous changes of potential energy as under :

region I : $x < 0$, $V(x) = 0.$

region II : $0 < x < L$, $V(x) = V_0.$

region III : $x > L$, $V(x) = 0.$

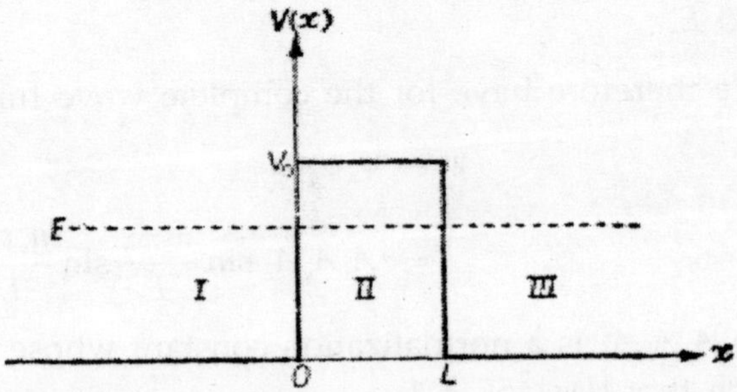

According to classical theory, if the particle's energy $E > V_0$, the particle will *always* pass the barrier. If $E < V_0$ (as in Fig.), then the particle will always be reflected back, it can never penetrate through the barrier.

The behaviour of the particle is, however, quite different according to wave mechanics. First, even when $E > V_0$, there is a finite probability that the particle will be reflected back from the barrier second when $E < V_0$, there is a finite probability that the particle will penetrate through" the barrier Higher the barrier, and the thicker it is, the smaller the chance the particle can penetrate it. Such a behaviour of the particle, absolutely impossible from the classical viewpoint, follows directly from the Schrodinger equation.

Let us consider the case $E < V_0$. The Schrodinger wave equations for the three regions have the following forms :

$$\frac{d^2\psi_1}{dx^2} + \frac{8\pi^2 m}{h^2} E\psi_1 = 0,$$

$$\frac{d^2\psi_2}{dx^2} + \frac{8\pi^2 m}{h^2}(E - V_0)\psi_2 = 0, \qquad (E < V_0)$$

and
$$\frac{d^2\psi_3}{dx^2} + \frac{8\pi^2 m}{h^2} E\psi_3 = 0.$$

Let us put $\alpha^2 = \dfrac{8\pi^2 mE}{h^2}$ and $\beta^2 = \dfrac{8\pi^2 m(V_0 - E)}{h^2}$.

The eq. become

$$\frac{d^2\psi_1}{dx^2} + \alpha^2\psi_1 = 0$$

$$\frac{d^2\psi_2}{dx^2} + \beta^2\psi_2 = 0$$

and
$$\frac{d^2\psi_3}{dx^2} + \alpha^2\psi_3 = 0.$$

The solutions to these equations are

$$\psi_1 = A_1 e^{\iota\alpha x} + B_1 e^{-\iota\alpha x}, \qquad \text{... } (i)$$

$$\psi_2 = A_2 e^{\iota\beta x} + B_2 e^{-\iota\beta x}, \qquad \text{... } (ii)$$

and
$$\psi_3 = A_3 e^{\iota\alpha x} + B_3 e^{-\iota\alpha x}, \qquad \text{... } (iii)$$

The constants may be interpreted as the amplitudes of the corresponding wave components, that is,

A_1 is the amplitude of the wave incident on the barrier at $x = 0$

B_1 is the amplitude of the wave reflected in region I

A_2 is the amplitude of the wave penetrating the barrier

B_2 is the amplitude of the wave reflected at $x = L$ in region II

A_3 is the amplitude of the wave transmitted in region III

B_3 is the amplitude of (non-existent) wave reflected in region III, and since there is no such wave, we must have

$$B_3 = 0.$$

To find the other constants, we use the conditions which the function ψ must satisfy. In order to be a proper wave function, ψ must be finite, single-valued and continuous everywhere along the x-axis. These conditions mean that at the boundaries $x = 0$ and $x = L$, the wave functions inside and outside must not only have the same value but also the same slope, so that they match up perfectly. Thus at the left-hand boundary, we must have

$$\psi_1 = \psi_2$$

$$\frac{d\psi_1}{dx} = \frac{d\psi_1}{dx} \qquad \text{at } x = 0$$

and at the right-hand boundary

$$\psi_2 = \psi_3$$

$$\frac{d\psi_2}{dx} = \frac{d\psi_3}{dx} \qquad \text{at } x = L.$$

These conditions when applied to eq. (*i*), (*ii*) and (*iii*) yield the following relations :

$$A_1 + B_1 = A_2 + B_2$$
$$i\alpha A_1 - i\alpha B_1 = \beta A_2 - \beta B_3$$
$$A_2 e^{\beta L} + B_2 e^{-\beta L} = A_3 e^{i\alpha L}$$
$$\beta A_2 e^{\beta L} - \beta B_2 e^{-\beta L} = i\alpha A_3 e^{t\alpha L}$$

Let us divide all the equations by A_1 and introduce the notations

$$\frac{B_1}{A_1} = b_1, \ \frac{A_2}{A_1} = a_2, \ \frac{B_2}{A_1} = b_3,$$

$$\frac{A_3}{A_1} = a_2 \text{ and } \frac{\beta}{\alpha} = n.$$

The above four equations then take the form

$$1 + b_1 = a_2 + b_2 \qquad \text{... } (iv)$$

$$i - ib_1 = na_2 - nb_2 \qquad \text{... } (v)$$

$$a_2 e^{\beta L} + b_2 e^{-\beta L} = a_2 e^{i\alpha L} \qquad \text{... } (vi)$$

$$na_2 e^{\beta L} - nb_2 e^{-\beta L} = ia_3 e^{i\alpha L} \qquad \text{... } (vii)$$

The ratio between the squares of the magnitudes of the transmitted and incident wave amplitudes

$$\mathrm{T} = \frac{|A_3|^2}{|A_1|^2} = |a_3|^2$$

determines the probability of a particle penetrating through the barrier and can be called the 'transmission coefficient'. To evaluate it, let us multiply (*iv*) by i and add it to (*v*), when we get

$$2i = (n + i)\, a_2 - (n - i)\, b_2. \qquad \text{... } (viii)$$

Now, multiplying (*vi*) by i and subtracting it from (*vii*), we get

$$(n - i)a_2 e^{\beta L} - (n + i)\, b_2 e^{-\beta L} = 0 \qquad \text{... } (ix)$$

Solving (*viii*) and (*ix*), we get

$$a_2 = \frac{2i(n+i)e^{-\beta L}}{(n+i)^2 e^{-\beta L} - (n-i)^2 e^{\beta L}}$$

and
$$b_2 = \frac{2i(n-i)e^{\beta L}}{(n+i)^2 e^{-\beta L} - (n-i)^2 e^{\beta L}}$$

Introducing these values of a_2 and b_2 in eq. (*vi*), we get

$$a_3 = \frac{4nie^{-i\alpha L}}{(n+i)^2 e^{-\beta L} - (n-i)^2 e^{\beta L}} \cdot$$

The quantity $\beta L = \frac{\sqrt{[2m(V_0 - E)]}}{h/2\pi} L$ is usually much greater than 1, and as such the term carrying $e^{-\beta L}$ in the denominator can be ignored.

Thus
$$a_3 \approx \frac{-4ni\, e^{-i\alpha L}}{(n-i)^2} e^{-\beta L}.$$

Its conjugate is
$$a_3{}^* \approx \frac{4ni\, e^{i\alpha L}}{(n+i)^2} e^{-\beta L}.$$

The square of the absolute value of a_2 is thus given by

But $|a_2|^2$ is the transmission coefficient or the penetration probability, T. Therefore

$$T \approx \frac{16\, n^2}{(n^2+1)^2} e^{-2\beta L}.$$

The expression $16n^2/(n^2+1)^2$ has a magnitude of the order of unity. We can therefore write

$$T \approx e^{-2\beta L}.$$

We had put
$$\beta^2 = \frac{8\pi^2 m(V_0 - E)}{h^2}.$$ Thus

This is the expression for the probability of penetration. It shows that the penetration probability rapidly falls off if the mass m of the particle, or the barrier width L, or the energy difference $V_0 - E$ increases. In fact, it is a phenomenon of the microscopic world.

We have reached the remarkable conclusion that *if a particle with energy E is incident on a thin energy barrier of height greater than E, there is a finite probability of the particle penetrating the barrier*. This phenomenon is called 'tunnel effect' because the particle crosses the barrier "without going over the top" as if it has passed through a "tunnel" in the barrier.

Tunnel effect is purely a quantum mechanical phenomenon, absolutely inconceivable in classical physics. The reason is that a particle "in the tunnel" ought to have a negative kinetic energy (in the tunnel $E < V_0$). In quantum mechanics, the division of the total energy into kinetic and potential energies has no sense because it contradicts the uncertainty principle. If we say that a particle has a definite kinetic energy K it would mean that it has a definite momentum. Similarly, a particle with a definite potential energy V would signify that it has a definite location in space. Since the momentum and location of a particle cannot simultaneously have definite values, it is impossible to find simultaneously exact values of kinetic and potential energies. Thus, although the total energy E of a particle has a quite definite value, it cannot be represented in the form of the sum of the exactly determined energies K and V. Hence in quantum mechanics the conclusion that K is negative inside the tunnel is meaningless.

Examples of Tunnel Effect

There are many examples in nature of quantum mechanical tunnelling. We consider a few of them.

1. *Alpha Decay:* The tunnel effect occurs in the emission of alpha particles from radioactive nuclei. The average

energy of an alpha particle formed within a nucleus is only a few MeV whereas the potential barrier around the nucleus (formed by the nuclear binding forces) is about 25 MeV high. Classically, the alpha particle cannot escape from the nucleus but quantum-mechanically it "tunnels" through the barrier. This tunnelling constitutes radioactive alpha decay.

2. *Field Emission:* The tunnel effect has also been found responsible for the "field emission" of electrons from metals. This is the emission of electrons by (cold) metals in vacuum subjected to strong electric fields. Classically, the field must be much more stronger than the actual fields at which the emission occurs. The actual field, however, reduces the width of the barrier for electrons at the metal-vacuum interface, so that electrons of energy less than the barrier height can "tunnel" through the barrier.

3. *Field Ionisation:* The tunnel effect also explains "field ionisation" of electrons inside solid insulators. In this phenomenon the electrons break away from their parent atoms under the influence of electric fields applied to insulators which thus undergo dielectric breakdown. Here, too, the actual field is weaker than classical physics requires it to be. But it makes the electrons to 'tunnel' through the potential barriers around the atoms.

4. *Tunnel Diode:* Crystal diodes are modified by an extremely thin layer of insulation between the two faces in contact. These are called "tunnel diodes" in which current flows by electrons tunnelling from one side to the other through potential barriers higher than the kinetic energy of the electrons. The rate of tunnelling, and therefore the current, can be regulated merely by changing the height of the barrier, which can be done with an applied voltage. This can be done rapidly, so

that frequencies higher than 10^9 Hz can be obtained. (Ordinary diodes depend on the diffusion of electrons across a junction and therefore operate at much lower frequencies.)

5. *Ammonia Inversion :* The figure below represents ammonia molecule NH_3. The coulomb repulsion of the three hydrogens creates a barrier against the nitrogen atom moving to a symmetric position (shown by dotted lines) on the opposite side of the plane of hydrogens. According to classical mechanics, we will have to give sufficient energy to the nitrogen atom before it can move along the axis of the molecule. Quantum-mechanically, the nitrogen atom can tunnel through the barrier and appear on the other side of the molecule. In fact, the N atom tunnels back and forth with a frequency $\approx 10^{10}$ Hz.

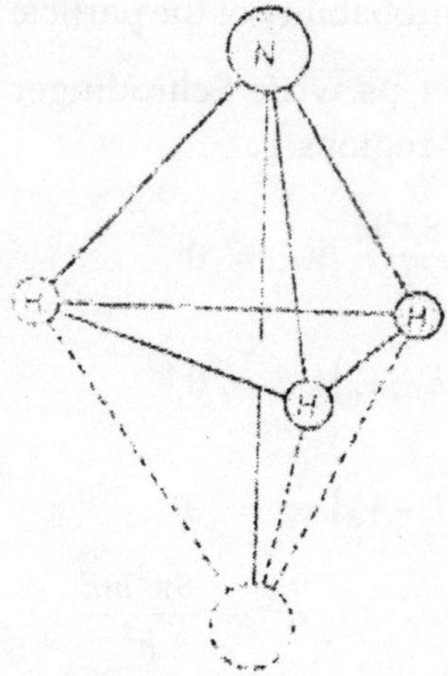

Particle in a Finite Square Potential Well (Non-rigid Box): Let us consider a particle with energy E moving along the x-axis in a square potential well of width L (Fig.). Suppose the potential energy of the particle is zero within the well, but rises suddenly to V_0 outside, where $V_0 > E$. We have three different regions separated by discontinuous changes of potential energy as under:

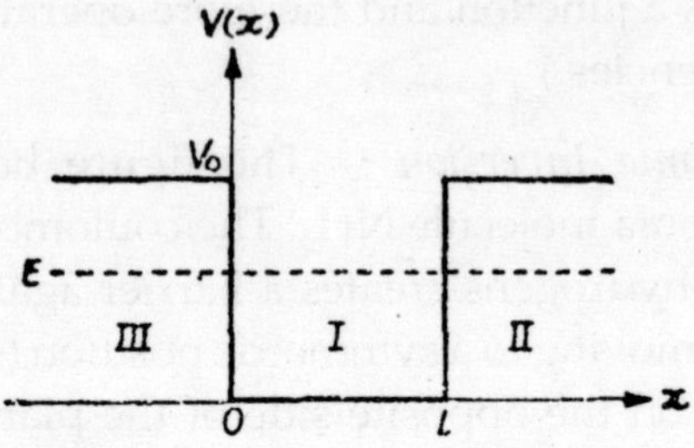

region I $\quad 0 < x < L,\ V(x) = 0$

region II $\quad x > L,\ V(x) = V_0$

region III $\quad x < 0,\ V(x) = V_0.$

Since the particle's energy E is smaller than the potential energy V_0 outside, classically the particle cannot break through the walls to come out. According to wave mechanics, however, there is a definite probability of the particle being found outside.

To see this, let us write Schrodinger wave equations for each of the three regions :

I. $$\frac{d^2\psi_1}{dx^2} + \frac{8\pi^2 m}{h^2} E\psi_1 = 0$$

II. $$\frac{d^2\psi_2}{dx^2} + \frac{8\pi^2 m}{h^2}(E - V_0)\,\psi_2 = 0$$

III. $$\frac{d^2\psi_3}{dx^2} + \frac{8\pi^2 m}{h^2}(E - V_0)\,\psi_3 = 0$$

Let us put $$\alpha^3 = \frac{8\pi^2 mE}{h^2}$$

and $$\beta^2 = \frac{8\pi^2 m(E - V_0)}{h^2}.$$

The eq. become

$$\frac{d^2\psi_1}{dx^2} + \alpha^2\psi_1 = 0$$

$$\frac{d^2\psi_2}{dx^2} + \beta^2\psi_2 = 0$$

$$\frac{d^2\psi_3}{dx^2} + \beta^2\psi_3 = 0$$

The solutions to these equations are

$$\psi_1 = A_1 \sin \alpha x + B_1 \cos \alpha x. \quad \text{... } (i)$$

$$\psi_2 = A_2 e^{\beta x} + B_2 e^{-\beta x}, \quad \text{... } (ii)$$

$$\psi_3 = A_3\, e^{\beta x} + B_3 e^{-\beta x}. \quad \text{... } (iii)$$

For ψ to remain finite as $x \to \pm\, \infty$, we must have

$$A_2 = 0 \text{ and } B_3 = 0.$$

Further, at the boundaries $x = 0$ and $x = L$, the wave functions inside and outside must have the same value and the same slope, *i.e.*, we must have

$$\psi_1 = \psi_3 \text{ and } \frac{d\psi_1}{dx} = \frac{d\psi_3}{dx} \text{ at } x = 0$$

and

$$\psi_1 = \psi_2 \text{ and } \frac{d\psi_1}{dx} = \frac{d\psi_2}{dx} \text{ at } x = L.$$

These conditions when applied to eq. (*i*), (*ii*) and (*iii*) yield the following relations :

$$B_1 = A_3 \quad \text{... } (iv)$$

$$\alpha A_1 = \beta A_3 \quad \text{... } (v)$$

$$A_1 \sin \alpha L + B_1 \cos \alpha L = B_2 e^{-\beta L} \quad \text{... } (vi)$$

$$\alpha A_1 \cos \alpha L - \alpha B_1 \sin \alpha L = -\beta\, B_2\, e^{-\beta L} \quad \text{... } (vii)$$

Eq. (*iv*) and (*v*) give

$$A_1 = \frac{\beta}{\alpha} B_1 .$$

This together with equations (*vi*) and (*vii*) yields

$$\tan \alpha L = \frac{2\alpha\beta}{\alpha^2 - \beta^2}$$

or $$\tan\left(\frac{2\pi L}{h}\sqrt{(2mE)}\right) = \frac{2\sqrt{[E(V_0 - E)]}}{2E - V_0}$$

Only those values of E which satisfy this relation are the allowed energy states. Thus the energy is quantised. The allowed energy values are found by numerical or graphical methods.

The wave functions for the first three allowed energy values and the corresponding probability densities for finding the particle at different locations are shown in figures below (*a*) and (*b*) respectively. It is seen that the wave functions ψ_1, ψ_2, ψ_3 do not vanish at the walls as they do for a particle in an infinite potential well. The particle wavelengths that can fit into the well are therefore somewhat longer and hence the allowed energies are lower, than those for a rigid box of infinitely deep well.

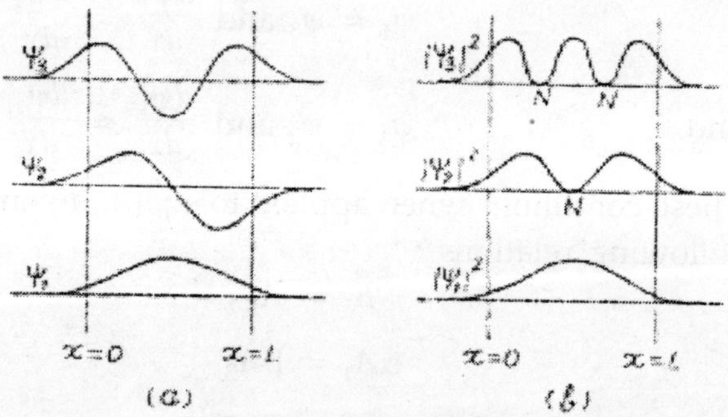

The probability density curves in figures above (*b*) show that in each energy state there is a finite probability of finding the particle outside the well although classically it can never be there. Such a probability is higher for higher energy state. Further, classically the particle moving with uniform velocity in the well would be equally likely to be found at all places. Wave mechanically the situation is quite different. In the lowest energy state the probability of the particle being found is maximum in the middle and decreases towards the walls. In higher energy states the maximum probability shifts towards the walls and there are points, marked N in the Fig., where there is zero probability of finding the particle.

The Harmonic Oscillator (Parabolic Potential Well): A simple harmonic oscillators a particle performing one-dimensional motion under a restoring force F which is proportional to the particle's displacement x from its equilibrium position, that is,

$$F = -kx,$$

where k is the force constant. The classical equation of motion of the particle is

$$m\frac{d^2x}{dt^2} = -kx$$

which yields a frequency given by

$$v = \frac{1}{2\pi}\sqrt{\left(\frac{k}{m}\right)}.$$

The potential energy of the particle has the form

$$V = \frac{1}{2}kx^2.$$

The plot of V against x is a parabola and we may describe the particle as being in a 'parabolic potential well. If E be the total energy of the parade (classically E can have any value), if oscillates back and forth between $x = -A$ and $x = +A$. The amplitude A is determined by the energy E, such that

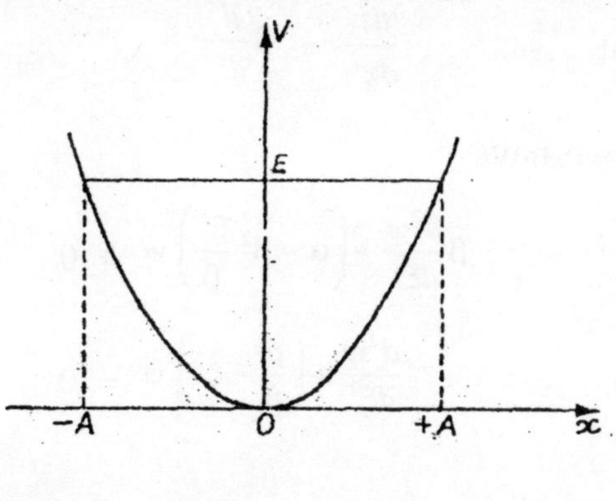

$$E = \frac{1}{2} kA^2$$

The velocity of the particle is maximum at its centre of path and drops to zero at the ends (at $x = \pm A$). Thus, classically, the probability of finding the particle is minimum at the centre (where it spends minimum time) and maximum at the ends (where it spends maximum time). Further, classically the particle cannot go beyond the limits of $-A$ and $+A$ because this would mean $V > E$ *i.e.* a negative kinetic energy.

Let us now consider a wave-mechanical harmonic oscillator.

The Schrodinger equation for the oscillator is, with $V = \frac{1}{2} kx^2$,

$$\frac{d^2\psi}{dx^2} + \frac{8\pi^2 m}{h^2}\left(E - \frac{1}{2} kx^2\right)\psi = 0$$

Putting $\alpha = \dfrac{8\pi^2 mE}{h^2}$ and $\beta = \sqrt{\left(\dfrac{4\pi^2 mk}{h^2}\right)}$, we get

$$\frac{d^2\psi}{dx^2} + (\alpha - \beta^2 x^2)\psi = 0$$

Let us introduce a dimensionless independent variable ξ given by

$$\xi = \sqrt{\beta} x$$

so that $$\frac{d^2}{dx^2} = \beta\frac{d^2}{d\xi^2}.$$

Then we have

$$\beta\frac{d^2\psi}{d\xi^2} + \left(\alpha - \beta^2\frac{\xi^2}{\beta}\right)\psi = 0$$

or $$\frac{d^2\psi}{d\xi^2} + \left(\frac{\alpha}{\beta} - \xi^2\right)\psi = 0 \qquad \text{... } (i)$$

Let us try a solution of this equation of the following form:

$$\psi(\xi) = CU(\xi)e^{-\xi^2/2} \qquad ...(ii)$$

Making this substitution in eq. (*i*), we get

$$\frac{d^2U(\xi)}{d\xi^2} - 2\xi\frac{dU(\xi)}{d\xi} + \left(\frac{\alpha}{\beta} - 1\right)U(\xi) = 0$$

If we replace $\frac{\alpha}{\beta} - 1$ by $2n$, then this equation becomes the Hermite differential equation. Hence we may replace $U(\xi)$ by $H_n(\xi)$. Then

$$\frac{d^2H_n(\xi)}{d\xi^2} - 2\xi\frac{dH_n(\xi)}{d\xi} + 2nH_n(\xi) = 0.$$

This shows that the solution of eq. (*i*) is obtained by replacing $U(\xi)$ by the Hermite polynomial $H_n(\xi)$ in eq. (*ii*). That is

$$\psi(\xi) = CH_n(\xi)e^{-\xi^2/2}.$$

This solution is known to be acceptable only for $n = 0$, 1, 2,...

The restriction on n gives a corresponding restriction on E. We have

$$\frac{\alpha}{\beta} = 2n + 1.$$

But $$\alpha = \frac{8\pi^2 mE}{h^2}$$

and $$\beta = \frac{2\pi\sqrt{(mk)}}{h}.$$

Therefore

$$\frac{8\pi^2 mE/h^2}{2\pi\sqrt{(mk)/h}} = 2n + 1 = 2\left(n + \frac{1}{2}\right).$$

or $$E = \left(n + \frac{1}{2}\right)\frac{h}{2\pi}\sqrt{\left(\frac{k}{m}\right)}.$$

But $\frac{1}{2\pi}\sqrt{\left(\frac{k}{m}\right)}$ = v. Therefore

$$E = (n + ½)\, hv.$$

Thus the Schrodinger's equation for the oscillator has finite, unambiguous and continuous solutions at values of E given by

$$\boxed{E_n = \left(n + \frac{1}{2}\right) hv}$$

where n = 0, 1, 2, ... Thus the wave mechanical harmonic oscillator can take only certain discrete energies separated by intervals of hv, where v is the classical frequency of oscillation and h is Planck's constant. The energy levels are equidistant unlike the energy levels of a particle in a box whose spacing diverges.

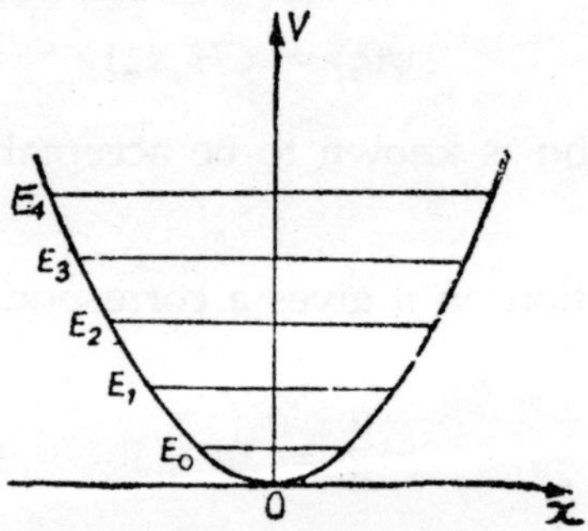

Further, for n = 0, the above expression gives

$$E_0 = ½\, hv,$$

which is the lowest possible value of the energy which the oscillator can have. This is called the 'zero-point energy'. Even if the temperature reduces to absolute zero, the oscillator would still have an amount of energy ½hv.

The existence of zero-point energy is confirmed by experiments on the scattering of light by crystals at low temperatures. It is found that with decreasing temperature the intensity of light scattered due to thermal oscillations of the atoms in a crystal lattice tends to a finite limit which remains unchanged with further decrease in temperature. This indicates that even at absolute zero the oscillations of the atom in the crystal do not stop.

It is also interesting to study the wave mechanical probability of finding the oscillator at different positions of its path in comparison with the classical probability. In figure below (*a*) is shown probability density curve for the lowest ($n = 0$) energy state. The classical probability density is shown by the dotted curve. Classically, the oscillating particle is most likely to be found at the ends of the path and least likely in the middle. Wave mechanically, the behaviour is just opposite. As shown, the probability density $|\psi_0|^2$ is maximum in the middle ($x = 0$) and drops off at either end. Further, wave mechanically the oscillator may be found beyond $x = \pm A$, as against the classical theory.

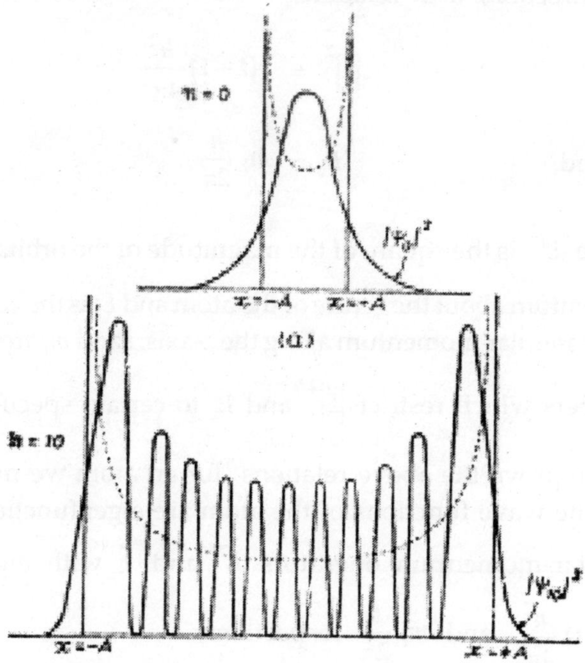

In figure above (*b*) is shown the probability density curve for a higher ($n = 10$) energy state. It is seen that the disagreement between wave mechanical and classical theories has reduced. The $|\psi_{10}|^2$ curve when averaged over x, approaches the classical probability curve (shown dotted). Also the "tails" of $|\psi_{10}|^2$ beyond $x = \pm A$ have decreased. This is an example of the Bohr's correspondence principle which states that in the limit of large quantum numbers the wave mechanics yields the same results as classical mechanics.

Angular Momentum

The angular momentum of atoms and molecules is almost as important as the energy as it helps to give physical significance to our mathematical functions. For a one-electron atom, the angular momentum $\vec{L}$ is quantised both in magnitude and direction, with relations

$$|\vec{L}|^2 = l(l+1)\frac{h^2}{4\pi^2}$$

and
$$L_z = m_l \frac{h}{2\pi},$$

where $|\vec{L}|^2$ is the square of the magnitude of the orbital angular momentum about the centre of the atom and L_z is the component of the angular momentum along the z-axis. l and m_l are quantum numbers which restrict $|\vec{L}|^2$ and L_z to certain specific values.

To prove the above relations for an atom we must show that the wave functions of the atom are eigenfunctions of the angular-momentum operators $\hat{L}^2$ and $\hat{L}_z$ with eigenvalues $l(l+1)\frac{h^2}{4\pi^2}$ and $m_l \frac{h}{2\pi}$ respectively.

The angular momentum $\vec{L}$ of a particle bound to and moving around a coordinate origin is defined by the equation

$$\vec{L} = \vec{r} \times \vec{p}$$

where $\vec{r}$ is the position vector of the particle relative to the origin and $\vec{p}$ is its linear momentum vector. The rectangular components of $\vec{L}$ are, from the definition of the cross product,

$$L_x = yp_z - zp_y$$

$$L_y = zp_x - xp_z$$

and $$L_z = xp_y, - yp_x ,$$

where x, y, z are the components of $\vec{r}$ and p_x, p_y, p_z are the components of $\vec{p}$.

Replacing p_x, p_y, p_z, by their equivalent differential operators $\frac{ih}{2\pi}\frac{\partial}{\partial x}, -\frac{ih}{2\pi}\frac{\partial}{\partial y}, -\frac{ih}{2\pi}\frac{\partial}{\partial z}$ respectively, we get the quantum mechanical operators for the rectangular components of angular momentum, which are

$$\hat{L}_x = -\frac{ih}{2\pi}\left(y\frac{\partial}{\partial z} - z\frac{\partial}{\partial y}\right)$$

$$\hat{L}_y = -\frac{ih}{2\pi}\left(x\frac{\partial}{\partial x} - x\frac{\partial}{\partial z}\right)$$

and $$\hat{L}_z = \frac{ih}{2\pi}\left(x\frac{\partial}{\partial y} - y\frac{\partial}{\partial x}\right).$$

In spherical polar coordinates these operators become*

$$\hat{L}_x = \frac{ih}{2\pi}\left(\sin\phi\frac{\partial}{\partial\theta} + \cot\theta\cos\phi\frac{\partial}{\partial\phi}\right) \quad \dots (i)$$

$$\hat{L}_y = \frac{ih}{2\pi}\left(-\cos\phi\frac{\partial}{\partial\theta} + \cot\theta\sin\phi\frac{\partial}{\partial\phi}\right) \quad \text{... (ii)}$$

and
$$\hat{L}_z = \frac{ih}{2\pi}\frac{\partial}{\partial\phi} \quad \text{... (iii)}$$

The square of the magnitude of $\vec{L}$ is

$$|\vec{L}|^2 = L_x^2 + L_y^2 + L_z^2$$

The corresponding operator is

$$\hat{L}^2 = \hat{L}_x^2 + \hat{L}_y^2 + \hat{L}_z^2$$

By substituting the squares of $\hat{L}_x, \hat{L}_y, \hat{L}_z$ (*i*), (*ii*), (*iii*), we obtain

$$\hat{L}_2 = -\frac{h^2}{4\pi^2}\left[\left\{\sin^2\phi\frac{\partial^2}{\partial\theta^2} + \cot^2\theta\cos^2\phi\frac{\partial^2}{\partial\phi^2}\right.\right.$$

$$\left. + \sin\phi\frac{\partial}{\partial\theta}\left(\cot\theta\cos\phi\frac{\partial}{\partial\phi}\right) + \cot\theta\cos\phi\frac{\partial}{\partial\theta}\left(\sin\phi\frac{\partial}{\partial\theta}\right)\right\}$$

$$+\left\{\cos^2\phi\frac{\partial^2}{\partial\theta^2} + \cot^2\theta\sin^2\phi\frac{\partial}{\partial\phi^2} - \cos\phi\frac{\partial}{\partial\theta}\left(\cot\theta\sin\phi\frac{\partial}{\partial\phi}\right)\right.$$

$$\left.\left. + \cot\theta\sin\phi\frac{\partial}{\partial\theta}\left(-\cos\phi\frac{\partial}{\partial\theta}\right)\right\} + \left\{\frac{\partial^2}{\partial\phi^2}\right\}\right]$$

$$-\cos\phi\frac{\partial}{\partial\theta}\left(\cot\theta\sin\phi\frac{\partial}{\partial\phi}\right)$$

$$\left. + \cot\theta\sin\phi\frac{\partial}{\partial\theta}\left(-\cos\phi\frac{\partial}{\partial\theta}\right)\right\} + \left\{\frac{\partial^2}{\partial\phi^2}\right\}\Bigg]$$

$$= \frac{h^2}{4\pi^2}\left[\frac{\partial^2}{\partial\theta^2} + \left(\cot^2 + 1\right)\frac{\partial^2}{\partial\phi^2}\sin\phi\left(-\text{cosec}^2\,\theta\right)\cos\phi\frac{\partial}{\partial\,\phi}\right.$$

$$\left. + \cos\theta\cos^2\phi\frac{\partial}{\partial\theta} - \cos\phi\left(-\text{cosec}^2\,\theta\right)\sin\phi\frac{\partial}{\partial\phi v} + \cot\theta\sin^2\phi\frac{\partial}{\partial\theta}\right]$$

$$= -\frac{h^2}{4\pi^2}\left[\frac{\partial^2}{\partial\theta^2} + \frac{1}{\sin^2\theta}\frac{\partial^2}{\partial\phi^2} + \cot\theta\,\frac{\partial}{\partial\theta}\right]$$

$$= -\frac{h^2}{4\pi^2}\left[\frac{\cos\theta}{\sin\theta}\frac{\partial}{\partial\theta} + \frac{\partial^2}{\partial\theta^2} + \frac{1}{\sin^2\theta}\frac{\partial^2}{\partial\phi^2}\right]$$

or $$\hat{L}^2 = -\frac{h^2}{4\pi^2}\left[\frac{1}{\sin\theta}\frac{\partial}{\partial\theta}\left(\sin\theta\frac{\partial}{\partial\theta}\right) + \frac{1}{\sin^2\theta}\frac{\partial^2}{\partial\phi^2}\right]. \quad \text{... } (iv)$$

This is the expression for the operator of the square of the angular momentum.

Let us now first apply the operator $\hat{L}_x$ (eq. *iii*) to the one electron atom wave function

$$\psi\ (\text{r}, \theta, \phi,) = R\ (r)\ \Theta\ (\theta)\ \Phi\ (\phi).$$

This gives

$$\hat{L}_z\ \psi = -\frac{ih}{2\pi}\frac{\partial\psi}{\partial\phi}$$

or $$L_z\ \psi = -\frac{ih}{2\pi} R\,\Theta\frac{\partial\Phi}{\partial\phi}.$$

The function Φ (ϕ) for the atom is given by

$$\Phi = Ae^{iml\psi}$$

Differentiating it,

$$\frac{d\Phi}{d\phi} = Aimle^{iml\psi} = imt\ \Phi.$$

Substituting it in the above expression for $\hat{L}_z\,\psi$, we get

$$\hat{L}_z\,\psi = -\frac{ih}{2\pi} im,\ R\ \Theta\ \Phi$$

or $$\boxed{\hat{L}_z\ \psi = mt\,\frac{h}{2\pi}\,\psi.}$$

$$[\ \therefore\ \psi = R\ \Theta\ \Phi]$$

This result means that the wave functions ψ of the one-electron atom are the eigenfunctions of $\hat{L}_z$ having eigenvalues given by

$$L_z = mt\ \frac{h}{2\pi}.$$

This is the expression for the quantised values of the z-component of the angular momentum if the atom. (L_x and L_y however do not obey quantisation relations).

We next apply the operator $\hat{L}^2$ (eq. iv) to the wave function $\psi = R\ \Theta\ \Phi$. We have

$$\hat{L}^2\ \psi\ = \frac{h^2}{4\pi^2}\left[\frac{1}{\sin\theta}\frac{\partial}{\partial\theta}\left(\sin\theta\frac{\partial}{\partial\theta}\right)+\frac{1}{\sin^2\theta}\frac{\partial}{\partial\phi^2}\right]R\ \Theta\ \Phi$$

$$= -\ \frac{h^2}{4\pi^2}\left[\frac{\Phi}{\sin\theta}\frac{d}{\partial\theta}\left(\sin\theta\frac{\partial\,\Theta}{d\theta}\right)+\frac{\Theta}{\sin^2\theta}\frac{d^2\Phi}{d\phi^2}\right].$$

Differentiating the Φ-equation

$$\Phi = Ae^{imi\psi}$$

twice, we get

$$\frac{d^2\Phi}{d\phi^2} = A(iml)^2\,e^{imi\psi} = -\ mr^2\ \Phi.$$

This gives

$$\hat{L}^2\ \psi = \frac{h^2}{4\pi^2}\ R\ \Phi\left[\frac{1}{\sin\theta}\frac{d}{d\theta}\left(\sin\theta\frac{d\Theta}{d\theta}\right)-\frac{ml^2}{\sin^2\theta}\Theta\right].$$

The function Θ (θ) is a solution of the equation

$$\frac{1}{\sin\theta}\frac{d}{d\theta}\left(\sin\theta\frac{d\Theta}{d\theta}\right)+\left[l\left(1+1\right)-\frac{ml^2}{\sin^2\theta}\right]\Theta = 0.$$

This means that the quantity in brackets in the above expression for $\hat{L}^2\ \psi$ is equal to $-\ l\ (l\ +\ 1)\ \Theta$. Hence we get

$$\hat{L}^2 \psi = -\frac{h^2}{4\pi^2} R \, \Phi \, [-l(l+1) \, \Theta]$$

$$\text{or } \hat{L}^2 \psi = l(l+1) \frac{h^2}{4\pi^2} R \, \Theta \, \Phi$$

or

$$\boxed{\hat{L}^2 \psi = l\left(l+1\right)\frac{h^2}{4\pi^2} \psi.}$$

Again, this result means that the wave function ψ of the one-electron atom are the eigenfunctions of $\hat{L}^2$ having eigenvalues given by

$$\left|\vec{L}\right|^2 = l\,(l+1)\frac{h^2}{4\pi^2}.$$

This is the expression for the quantised values of the square of the angular momentum of the atom.

PROBLEMS

1. Let's see what is the lowest energy that a neutron (mass 1.67 × 10^{-27} kg) can have if confined to move along the edge of an impenetrable box of length 10^{-14} meter (h = 6.63 × 10^{-31} joule-sec).

Solution : The quantised energies of a particle of mass m in a one-dimensional box of length L are given by

$$E_n = \frac{n^2 h^2}{8mL^2} \qquad n = 1, 2, 3, \ldots$$

The lowest energy corresponds to $n = 1$, and is thus

$$E_1 = \frac{h^2}{8mL^2}$$

Substituting the given values, we have

$$E_1 = \frac{\left(6.63 \times 10^{-34} \text{ joule - sec}\right)^2}{8 \times \left(1.67 \times 10^{-27} \text{ kg}\right) \times \left(10^{-14} \text{ meter}\right)^2}$$

$$= 3.29 \times 10^{-13} \text{ joule}$$

$$= \frac{3.19 \times 10^{-13}}{1.6 \times 10^{-13}}$$

$$(\therefore 1 \text{ Me V} = 1.6 \times 10^{-13} \text{ joule})$$

$$= 2.05 \text{ MeV}.$$

2. *An electron is confined to move between two rigid walls separated by 10^{-9} meter. Let's find the de Broglie wavelengths representing the first three allowed energy states of the electron and the corresponding energies. (electron mass is 9.1×10^{-31} kg and $h = 6.63 \times 10^{-34}$ joule-sec).*

Solution : The electron moving back and forth between rigid walls will form a stationary wave-pattern with nodes at the walls. For this, the distance L between the walls must be a whole multiple of the de-Broglie half-wave-lengths. Thus

$$L = n\frac{\lambda}{2}, \qquad n = 1, 2, 3, ...$$

or

$$\lambda = \frac{2L}{n}.$$

Here

$$L = 10^{-9} \text{ meter}$$

$$= 10 \text{ Å}.$$

$$\therefore \qquad \lambda = \frac{2 \times 10 \text{ Å}}{n} \qquad n = 1, 2, 3, ...$$

$$= 20 \text{ Å}, 10 \text{ Å}, 6.7 \text{ Å}, ...$$

The corresponding energies are given by

$$E_n = \frac{n^2 h^2}{8mL^2}.$$

Here

$$m = 9.1 \times 10^{-31} \text{ kg and } L = 10^{-9} \text{ meter}.$$

$$\therefore \qquad E_n = \frac{(6.63 \times 10^{-34})^2 n^2}{8 \times (9.1 \times 10^{31}) \times (10^{-9})^2}$$

$$= 6.04 \times 10^{-20}\ n^2 \text{ joule}$$

$$= \frac{6.40 \times 10^{-20}\ n^2}{1.6 \times 10^{-19}}$$

$$= 0.38\ n^2 \text{ eV.}$$

For $n = 1, 2, 3$, we have

$$E_1 = 0.38 \text{ eV},$$

$$E_2 = 1.52 \text{ eV},$$

$$E_3 = 3.42 \text{ eV}.$$

3. The energy difference between the ground state and the first excited state for an electron in a one-dimensional rigid box of length 10^{-8} cm is calculated as follows. (mass of electron = 9.1 × 10^{-31} kg and h — 6.626 × 10^{-34} J-S).

Solution : The energy of a particle of mass dimensional rigid box of side L is given by

$$E_n = \frac{n^2h^2}{8mL^2} \qquad n = 1, 2, 3, \ldots \ldots$$

$$= n^2 \frac{\left(6.626 \times 10^{-34}\ J-s\right)^2}{8 \times \left(9.1 \times 10^{-31}\ \text{kg}\right) \times \left(10^{-10}\ \text{m}\right)^2}$$

$$= 0.60 \times 10^{-17}\ n^2 \text{ J}$$

$$= \frac{0.60 \times 10^{-17}}{1.6 \times 10^{19}}\ n^2$$

$$= 38\ n^2 \text{ eV.}$$

For the ground state ($n = 1$) ; $E_1 = 38$ eV and for the first excited state ($n = 2$) ; E_2 — 152 eV. The energy difference is 152–38 = 114 eV.

4. The energy states for a ball of mass 10 gm moving in a box of length 10 cm. (h = 6.6 × 10^{-34} J-s) can not be observed.

Solution : On calculating as above, we shall have

$$E_n = 3.4 \times 10^{-45}\, n^2 \text{ eV}.$$

The first few (n = 1,2,3,...) energies will be 3.4×10^{-45} eV, 13.6×10^{-45} eV, 30.6×10^{-45} eV...... These are so near to each other that a continuum will be observed.

5. The probabilities of finding a particle trapped in a box of length L in the region from 0.45L to 0.55L for the ground state and the first excited state is found as follows.

Solution : The eigenfunctions of a particle trapped in a box of length L are

$$\psi_x(x) = \sqrt{\left(\frac{2}{L}\right)}\ \sin \frac{n\pi x}{L}.$$

The probability of finding the particle between x_1 and x_2 , when; it is in the nth state , is

$$P = \int_{x_1}^{x_2} |\psi_n|^2\, dx$$

$$= \frac{2}{L}\int_{x_1}^{x_2} \sin^2 \frac{n\pi x}{L}\, dx$$

$$= \frac{2}{L}\int_{x_1}^{x_2} \frac{1}{2}\left(1 - \cos \frac{2n\pi x}{L}\right) dx$$

$$= \frac{1}{L}\left[x - \frac{L}{2n\pi} \sin \frac{2n\pi x}{L}\right]_{x_1}^{x_2}.$$

Here

x_1 = 0.45L,

x_2 = 0 .55 L and for the ground state;

n = 1.

$$\therefore \quad P = \frac{1}{L}\left[x - \frac{L}{2\pi}\sin\frac{2\pi x}{L}\right]_{0.45L}^{0.55L}$$

$$= \frac{1}{L}\left[\left(0.55 - \frac{L}{2\pi}\sin 1.10\,\pi\right) - \left(0.45L - \frac{L}{2\pi}\sin 0.90\,\pi\right)\right]$$

$$= \left[\left(0.55 - \frac{L}{2\pi}\sin 198^\circ\right) - \left(0.45 - \frac{1}{2\pi}\sin 162^\circ\right)\right]$$

$$= (0.55 - 0.45) - \frac{1}{2\pi}\,(\sin 190^\circ - \sin 162^\circ)$$

$$= 0.10 - \frac{1}{\pi}\,(\cos 180^\circ \sin 18^\circ)$$

$$= 0.10 + \frac{0.3090}{3.14}$$

$$= 0.198 = 19.8\%.$$

For the first excited state ($n = 2$), we shall have

$P = 0.65\%$ (Try).

6. Electrons with energy of 1 eV are incident on a barrier 5 eV high and 5Å wide. Let's find the probability for these electrons to penetrate the barrier. The way it is affected if the barrier is doubled in width is explained here. Mass of electron is 9.1 × 10^{-31} kg and h 6.62 × 10^{-34} J-s.

Solution : The (approximate) transmission probability is

$$T = e^{-2BL}$$

$$\text{where } \beta = \frac{2\pi\sqrt{[2m\,(v_o - E)]}}{h}$$

For the 1-eV electron, we have

$$\beta = \frac{2 \times 3.14\sqrt{\left[2\left(9.1 \times 10^{31}\text{ kg}\right)(5-1)\text{eV}\left(1.6 \times 10^{19}\text{ j/eV}\right)\right]}}{6.62 \times 10^{-34}\text{ J-s}}$$

$$= 1.0 \times 10^{10}\text{ m}^{-1}$$

Now,

$$L = 5\text{Å} = 5 \times 10^{-10} \text{ m} \quad \text{and so}$$

$$2\beta L = 2 \times (1.0 \times 10^{10} \text{ m}^{-1}) \times (5 \times 10^{-10} \text{ m}) = 10.$$

$$\therefore T\, e^{-10} = (2.72)^{-10} = (0.368)^{10}$$

$$= 4.5 \times 10^{-5}.$$

One 1-eV electron out of 22,000 can tunnel through the 5-eV barrier on the average.

If the barrier is doubled in width to 10Å, the transmission probability becomes

$$T = e^{-20} = (0.368)^{20}$$

$$= 2.1 \times 10^{-9}.$$

6

Matter Waves

de-Broglie Hypothesis of Matter Waves: The wave theory of light successfully explained the phenomena such as interference and diffraction. It, however, failed to explain the photoelectric effect, the Compton effect, the anomalous Zeeman effect, the short wavelength limit of the continuous X-ray spectrum, etc. These phenomena have, however, been explained on the basis of Planck's quantum theory. According to quantum theory, a beam of light of frequency v is composed of tiny discrete packets called 'photons', each having an energy hv, where h is Planck's constant. Thus light possesses a dual character, behaving as a wave as well as a particle (photon).

In analogy with dual character of light, de-Broglie in 1924 introduced the hypothesis that *all material particles in motion possess a wave character also.* Accordingly, particles such as electrons, protons, etc. have waves associated with them. These waves are called 'matter waves' or 'de-Broglie waves'.

de Broglie Wavelength of Matter Waves: A light wave of frequency v is associated with a photon of energy E, given by

$$E = hv.$$

According to Einstein's theory of relativity, a mass m is equivalent to an amount of energy E, according to the relation

$$E = mc^2$$

or $$m = \frac{E}{c^2}$$

where c is the velocity of light in free space. For a photon $E = h\nu$. Hence the mass of the photon is given by

$$m = \frac{h\nu}{c^2}$$

The photon travels with the velocity of light, c. Therefore, the momentum of the photon is given by

$$p = mc$$

$$= \frac{h\nu}{c^2}c = \frac{h\nu}{c}.$$

But $\nu = \frac{c}{\lambda}$, where λ is the wavelength of light

$$\therefore \qquad p = \frac{h}{\lambda}$$

or $$\lambda = \frac{h}{p}.$$

Thus, *the wavelength associated with a photon is h/p.*

de-Broglie postulated that the wavelength associated with a moving material particle is, in analogy with light, given by

$$\lambda = \frac{h}{p}$$

where p is the momentum of the particle. A particle of (relativistic) mass m and moving with a velocity v has a momentum mv. Therefore, the wavelength associated with the moving particle is given by

$$\lambda = \frac{h}{p} = \frac{h}{mv}$$

de-Broglie Wavelength of Electron (Non-relativistic Case): Let us consider the case of electrons which have been accelerated from rest through a potential difference V. The velocity v acquired by these electrons (in the non-relativistic case) is given by

$$\frac{1}{2} m_0 v^2 = eV,$$

where m_0 is the rest mass and e the charge of the electron.

$$\therefore \qquad v = \sqrt{\left(\frac{2eV}{m_0}\right)}.$$

Now, the de-Broglie expression for the wavelength is

$$\lambda = \frac{h}{mv}.$$

If the relativistic variation of mass with velocity is to be ignored then

$$m \approx m_0$$

so that

$$\lambda = \frac{h}{m_0 v}$$

Substituting the value of v from above, we get

$$\lambda = \frac{h}{m_0 \sqrt{\left(\frac{2eV}{m_0}\right)}}$$

or

$$\lambda = \frac{h}{\sqrt{(2m_0 eV)}}.$$

This is the expression for the wavelength of an accelerated electron in the non-relativistic case.

Substituting $h = 6.625 \times 10^{-34}$ joule-sec, $m_0 = 91 \times 10^{-31}$ kg and $e = 16 \times 10^{-10}$ coulomb, we get

$$\lambda = \frac{6.625 \times 10^{-34}}{\sqrt{\left[2 \times \left(9.1 \times 10^{-31}\right) \times \left(1.6 \times 10^{-19}\right)\right]}} \frac{1}{\sqrt{V}}$$

$$= \frac{12.27 \times 10^{-10}}{\sqrt{V}} \text{ meter}$$

$$= \frac{12.27}{\sqrt{V}} \text{ Å} \simeq \sqrt{\left(\frac{150}{\sqrt{V}}\right)} \text{ Å}.$$

Demonstration of Matter Waves

Davisson and Germer Experiment on Electron Diffraction: If material particles have a wave character, they are expected to show phenomena like interference and diffraction. In 1927, Davisson and Germer demonstrated that a beam of electrons does suffer diffraction. Their apparatus is shown in Fig. Electrons from a heated filament are accelerated through a variable potential V and emerge from the 'electron gun' G. This electron beam falls normally on a nickel crystal C. The electrons are diffracted from the crystal in all directions. The intensity of the diffracted beam in different directions is measured by a Faraday cylinder F (connected to a galvanometer), which can be moved along a circular scale S. The crystal can be turned about an axis parallel to the incident beam. Thus any azimuth of the crystal can be presented to the plane defined by the incident beam and the beam entering the Faraday cylinder.

First of all, the accelerating potential V is given a low value, and the crystal is turned at any arbitrary azimuth. The faraday cylinder is moved to various positions on the scale S and the galvanometer current at each position is noted. The current, which is a measure of the intensity of the diffracted beam, is plotted against the angle ϕ between the incident beam, and the beam entering the cylinder. The observations are repeated for different accelerating potentials and the corresponding curves are drawn as shown in Fig. below.

It is seen that a 'bump' begins to appear in the curve for 44-volt electrons. With increasing potential the bump moves upwards and becomes most prominent in the curve for 54-volt electrons at $\phi = 50°$. At higher potentials the bump gradually disappears.

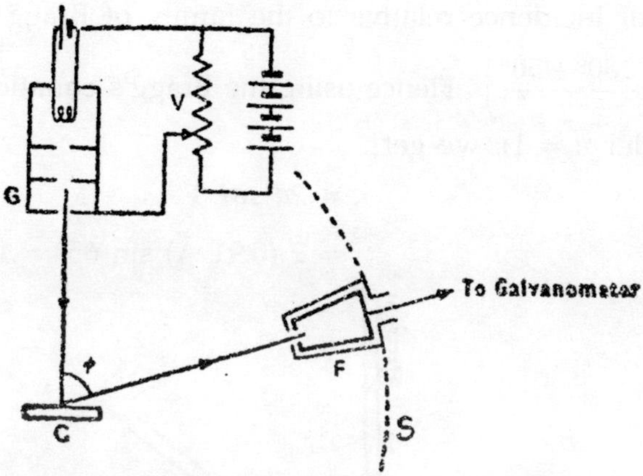

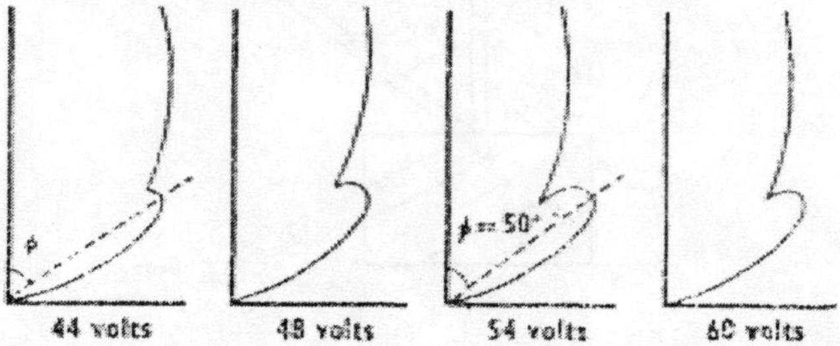

The bump in its most prominent state verifies the existence of electron waves. For, according to de Broglie, the wavelength associated with an electron accelerated through V volts is

$$\lambda = \frac{12.27}{\sqrt{V}} \text{ Å}$$

Hence, the wavelength associated with a 54-volt electron is

$$\lambda = \frac{12.27}{\sqrt{54}} = 1.67 \text{ Å}$$

Now, it is known from X-ray analysis that the nickel crystal acts as a plane diffraction grating with grating space d = 0.9 Å (Fig. below). According to experiment, we have a diffracted electron beam at $\phi = 50°$. This arises from wave-like diffraction from the family of Bragg atomic planes. The corresponding angle of incidence relative to the family of Bragg planes is $\theta = \left(= \dfrac{180° - 50°}{2}\right)$. Hence using the Bragg's equation (taking the order n = 1), we get

$$\lambda = 2d \sin \theta$$

$$= 2\ (0.91\ \text{Å}) \sin 65° = 1.65\ \text{Å}.$$

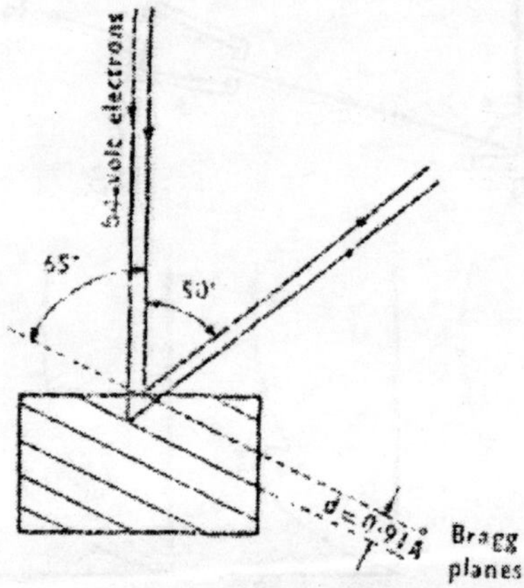

This, being in excellent agreement with the wavelength computed from de Broglie hypothesis, shows that electrons are wave-like in some circumstances. Other fundamental particles like neutrons also show wave-like properties. The Davisson-Germer experiment thus provides direct verification of de Broglie hypothesis of the wave nature of moving particles.

Electron Microscope

Conception of Matter Waves Trivial Under Ordinary Situations: The wave nature of matter is not apparent to our daily observations. The reason is that in our daily life we come across microscopic objects for which the de Broglie wavelength is extremely small. For example, the de Broglie wavelength of a bullet of mass 1 gm moving with a velocity of 10^5 cm/sec is given by

$$\lambda = \frac{h}{mv} = \frac{6.625 \times 10^{-34} \text{ joule-sec}}{10^{-3} \text{ kg} \times 10^{3} \text{ meter/sec}}$$

$$= 6.625 \times 10^{-34} \text{ meter}$$

This wavelength is too small to show any effect.

G. P. Thomson's Experiment

His arrangement is shown in Fig. The electrons are produced from a heated filament F and accelerated through a *high* positive potential given to the anode A. The whole apparatus is kept highly evacuated. The electron beam passes through a fine hole in a metal block B and falls on a gold foil G of thickness 10^{-8} meter. The electrons passing through the foil are received on a photographic plate P. The foil consists of very large numbers of microscopic crystals oriented at random. Therefore, some of them are always at the correct angles to give rise to diffraction according to Bragg's formula $2d \sin\theta = n\lambda$. Hence, on processing the plate, a central spot (marking the position of undeflected electrons) surrounded by a series of concentric diffraction rings is obtained. The diameters of these rings are measured.

Let AB (Fig. below) be a beam which has undergone a Bragg's diffraction by a small crystal in the gold foil G and falls on a photographic plate at a point E at a distance r from the central point C. The incident and the first-order diffracted rays make equal angles, θ with Bragg plane. Let $BC = L$.

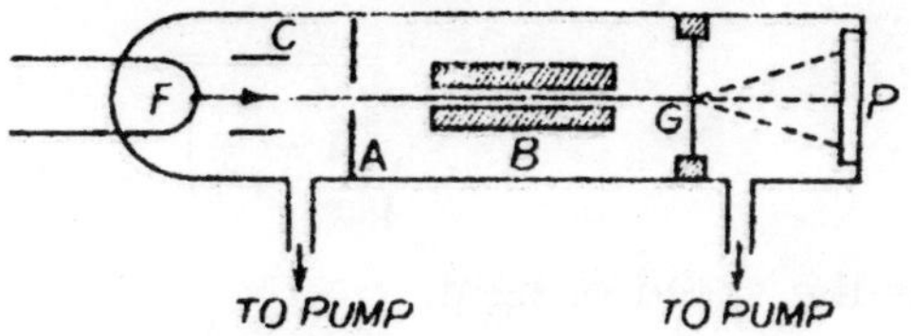

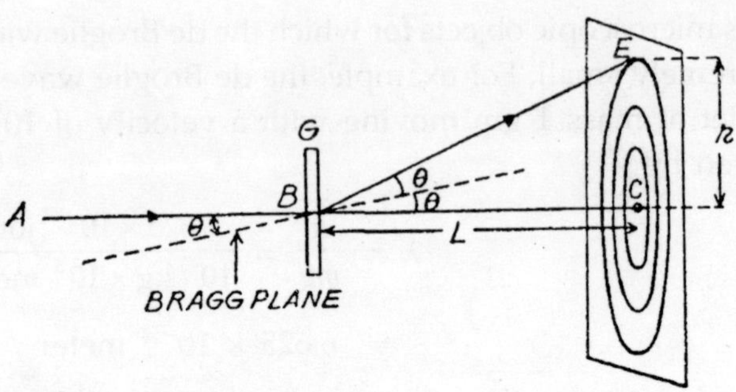

From Bragg's formula, we have

$$2d \sin \theta = n\lambda,$$

where d is the crystal lattice spacing. Therefore, for the first order ($n = 1$), we have

$$d = \frac{\lambda}{2 \sin \theta} = \frac{\lambda}{2\theta} \quad \text{(as } \theta \text{ is small)}$$

But, from the figure, $2\theta = \tan 2\theta = \dfrac{r}{L}$.

$$\therefore \quad d = \frac{L}{r}\lambda. \qquad \text{... } (i)$$

Now, the de Broglie wavelength associated with a moving particle is given by

$$\lambda = \frac{h}{mv},$$

where m is the (relativistic) mass and v the velocity of the particle. In Thomson's experiment the particles (electrons) are accelerated through a p.d. of about 15-60 kilovolts. For such energetic electrons the relativistic mass is to be distinguished from the rest mass m_0; and we have

$$m = \frac{m_0}{\sqrt{\left(1 - \frac{v^2}{c^2}\right)}},$$

where c is the speed of light.

$$\therefore \qquad \lambda = \frac{h}{mv} = \frac{h}{m_0 v \sqrt{\left(1 - \frac{v^2}{c^2}\right)}} \qquad \text{... } (ii)$$

The relativistic expression for the kinetic energy is

$$m_0 c^2 \left\{ \left(1 - \frac{v^2}{c^2}\right)^{-\frac{1}{2}} - 1 \right\}.$$

Therefore, if the electron (charge e) is accelerated from rest through a potential of V volts, we have

$$m_0 c^2 \left\{ \left(1 - \frac{v^2}{c^2}\right)^{-\frac{1}{2}} - 1 \right\} = eV$$

or
$$\left(1 - \frac{v^2}{c^2}\right)^{-1/2} = 1 + \frac{eV}{m_0 c^2} = 1 + z, \qquad \text{... } (iii)$$

where $z = eV/m_0c^2$.

$$\therefore \qquad 1 - \frac{v^2}{c^2} = (1 + z)^2$$

or
$$v = c[1 - (1 + z)^{-2}]^{1/2} \qquad \text{... } (iv)$$

Multiplying eq. (*iii*) and (*iv*), we get

$$\frac{v}{\sqrt{\left(1 - \frac{v^2}{c^2}\right)}} = c[1 - (1 + z)^{-2}]^{1/2}\,(1 + z)$$

$$= c[(1 + z)^2 - 1]^{1/2}$$

$$= c\left[2z\left(1 + \frac{z}{2}\right)\right]^{1/2}$$

$$= c\left[\frac{2eV}{m_0 c^2}\left(1 + \frac{eV}{2m_0 c^2}\right)\right]^{1/2}.$$

Substituting this value in eq. (*ii*), we obtain

$$\lambda = \frac{h}{m_0 c \sqrt{\left(\frac{2eV}{m_0 c^2}\right)}} \left(1 + \frac{eV}{2m_0 c^2}\right)^{-1/2}$$

or

$$\boxed{\lambda = \frac{h}{\sqrt{(2m_0 eV)}} \left(1 + \frac{eV}{2m_0 c^2}\right)^{-1/2}}$$

This is the relativistic expression for the de Broglie wavelength of an electron accelerated through a *high* p.d. of V volts. It reduces to the expression $\lambda = \frac{h}{\sqrt{(2m_0 eV)}}$, if relativistic effect is ignored (put $c = \infty$).

Substituting the above value of λ in eq. (*i*), we get

$$d = \frac{L}{r} \frac{h}{\sqrt{(2m_0 eV)}} \left(1 + \frac{eV}{2m_0 c^2}\right)^{-1/2}$$

Comparison of the value of d obtained in this way agreed to within 1% with values determined by using X-rays of known wavelengths. For example, in case of gold foil the value obtained by the above formula was 4.08 Å while that obtained by X-ray method was 4.06 Å. Thus the de Broglie conception of matter waves was verified.

de-Broglie Wavelength of Helium Atoms: The de-Broglie wavelength of a particle of mass m moving with a velocity v is given by

$$\lambda = \frac{h}{mv},$$

where h is Planck's constant. Let us consider a beam of the atoms at a temperature of T K. The atoms have a Maxwellian distribution of velocities. The mean energy of the atoms is given by

$$\frac{1}{2} m\overline{v^2} = \frac{3}{2} kT,$$

where k is Boltzmann's constant. Their root-mean-square velocity is

$$u_{rms} = \sqrt{(\overline{v^2})} = \sqrt{\left(\frac{3kT}{m}\right)}.$$

The de-Broglie wavelength of the atoms is therefore

$$\lambda = \frac{h}{mv_{rms}} = \frac{h}{m}\sqrt{\left(\frac{m}{3kT}\right)}$$

$$= \frac{h}{\sqrt{(3mkT)}}.$$

This is the required expression.

Experimental Verification: The wave nature of moving Hθ atoms was experimentally demonstrated by Estermann ; Frisch and Stern in 1930. They filled helium gas in an enclosure heated to 400 K, from which a beam of atoms emerged through a slit (Fig. below). In order to secure a beam of more uniform velocity, a device known as 'velocity-selector' was used. It consisted of a pair of parallel discs W_1 and W_2 mounted on a common shaft and rotated with a high velocity. The discs had slots cut in it in positions displaced relative to each other. Of the atoms passing through the slot in W_1, only those can get through the slot in W_2 whose velocity is such that by the time they arrive at W_2, the slot comes at the correct position.

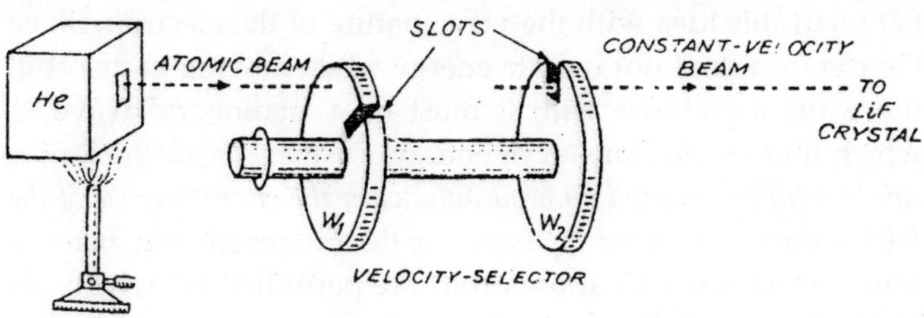

The beam emerging from W_2 was allowed to fall on the surface of a LiF crystal, and the atoms diffracted in a given direction were collected in a small evacuated chamber. The resulting helium gas pressure in this chamber was determined by means of a hot-wire gauge, which was a measure of the intensity of the diffracted beam. Thus the angle θ corresponding to the maximum intensity was known. From this, using the Bragg's equation

$$2d \sin \theta = n\lambda,$$

the wavelength of the beam was determined. This was found in very close agreement with the value calculated from the theoretical expression $\lambda = \dfrac{h}{\sqrt{(3mkT)}}$, where m is the mass of helium atom and T the temperature of the gas in the enclosure.

Bohr Quantisation Condition

According to de Broglie conception, an electron of mass m moving with velocity v has associated with it a wave whose wavelength is $\lambda = h/mv$, where h is the Planck's constant. This wave-character of electron can account for the limited number of permissible orbits in an atom and also enables us to deduce the quantised angular momentum of an electron in an atom.

We know that according to Bohr, the electron in an atom revolves in circular non-radiating orbits around the nucleus. Let us fit this idea with the wave-nature of the electron. Since the electron does not radiate energy while moving in its orbit, the wave associated with it must be a 'stationary' wave in which there is no transfer of energy. Further, *the electron wave can be adjusted around an orbit only when the circumference of the orbit is equal to an integral multiple of the wavelength* (Fig. below). This means that only those orbits are permitted for which the radius r is given by

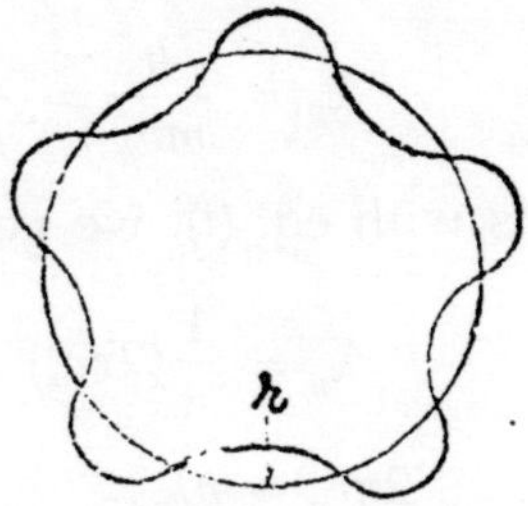

$$2\pi r = n\lambda = n\frac{h}{mv}$$

Hence the angular momentum of the electron in its orbit must be

$$L = mvr = \frac{nh}{2\pi},$$

which is the Bohr's quantisation condition.

de Broglie Wavelength and Circumference of Electron Orbits: From Bohr's theory, the radius of the nth orbit is given by

$$r_n = \frac{n^2h^2\varepsilon_0}{\pi me^2},$$

so that the circumference is

$$2\pi r_n = \frac{2n^2h^2\varepsilon_0}{me^2} \qquad \text{... } (i)$$

The velocity of the electron is given by

$$v_n = \frac{e^2}{2nh\varepsilon_0}.$$

The linear momentum of the electron is

$$mv_n = \frac{me^2}{2nh\varepsilon_0}$$

The de-Broglie wavelength of the electron in the *n*th orbit is

$$\lambda_n = \frac{h}{mv_n} = \frac{h}{me^2 / 2nh\varepsilon_0} = \frac{2nh^2\varepsilon_0}{me^2}.$$

Comparing this with eq. (*i*), we get

$$\lambda_n = \frac{1}{n}(2\pi r_n)$$

or $$2\pi r_n = n\lambda_n.$$

For $n = 1$; $$2\pi r_1 = \lambda_1$$

For $n = 2$; $$2\pi r_2 = 2\lambda_2,$$

and so on. Thus the circumference of the first Bohr orbit is equal to 1 complete wavelength in that orbit, that of the second orbit is equal to 2 complete wavelengths in that orbit, and so on.

Possible Energy States of a Particle in a Box: Let us consider a particle of mass m enclosed in a box of length L with impenetrable walls. Suppose the (bound) particle is moving back and forth in the x-direction with constant speed v, making perfectly elastic collisions with the walls of the box.

According to the de Broglie hypothesis, the particle has a wave associated with it whose wavelength λ is given by

$$\lambda = \frac{h}{p}$$

where p is the momentum of the particle. Since the walls of the box are impenetrable, the particle cannot move beyond the walls and so the amplitude of the associated wave must drop to zero at the walls. In other words, the particle moving back and forth between opposite walls will form a stationary-wave pattern with nodes at the walls. For this to happen, the distance L between the opposite walls must be a whole multiple of half-wavelengths. That is

or $$L = n\frac{\lambda}{2}, \qquad \text{where } n = 1, 2, 3, ...$$

The possible values of the momentum of the particle are therefore

$$p = \frac{h}{\lambda} = n\frac{h}{2L}.$$

Consequently, the possible values of the kinetic energy of the particle are

$$K = \frac{p^2}{2m} = n^2 \frac{h^2}{8mL^2}, \quad n = 1, 2, 3, ...$$

Thus only certain discrete energy states are possible for the particle bound in a box. The minimum kinetic energy (for $n = 1$) is, $h^2/8mL^2$.

Quantisation is not Noticed for Macroscopic Objects: There is a very good reason why with the objects of every-day life the quantum effect is not observed. Let us consider a ball of mass $m = 10$ gm (= 0.01 kg) moving in a box of length $L = 0.1$ meter. The quantum of energy for this ball would be

$$K_1 = \frac{h^2}{8mL^2}$$

$$= \frac{(6.62 \times 10^{-34} \text{ joule - sec})^2}{8 \times (0.01 \text{ kg}) \times (0.1 \text{ meter})^2}$$

$$= 5.5 \times 10^{-64} \text{ joule.}$$

The corresponding velocity would be

$$v = \sqrt{\left(\frac{2K_1}{m}\right)}$$

$$= \sqrt{\left(\frac{2 \times (5.5 \times 10^{-64} \text{ joule})}{0.01 \text{ kg}}\right)}$$

$$= 3.3 \times 10^{-31} \text{ meter/sec.}$$

Thus, unless the ball were given an unreasonably small velocity, the wavelength would be so small that no effects of quantisation of motion would be noticeable.

Wave Velocity

When a monochromatic wave (wave of a single frequency and wavelength) travels through a medium, its velocity of advancement in the medium is called the 'wave velocity'. For example, a plane harmonic wave travelling along the $+x$ direction is given by

$$y = a \sin (\omega t - kx),$$

where a is the amplitude, $\omega(= 2\pi n)$ is the angular frequency and $k(= 2\pi/\lambda)$ is the propagation constant of the wave. By definition, the ratio of angular frequency ω to the propagation constant k is the wave velocity v_p. That is,

$$\boxed{v_p = \frac{\omega}{k}}$$

$(\omega t - kx)$ is the phase of the wave-motion. Therefore, the planes of constant phase (wavefronts) are defined by

$$\omega t - kx = \text{constant}.$$

Differentiating with respect to time, we get

$$\omega - k\frac{dx}{dt} = 0$$

or

$$\frac{dx}{dt} = \frac{\omega}{k}$$

which is the wave velocity v_p. Thus *the wave velocity is the velocity with which planes of constant phase advance through the medium.* Hence the wave velocity is also called as 'phase velocity'.

Group Velocity

In practice, we come across pulses rather than monochromatic waves. A pulse consists of a number of waves differing slightly from one another in frequency. A superposition of these waves is called a 'wave packet' or a 'wave group'. When such a group travels in a medium, the phase velocities of its different components are different. The observed velocity is, however, the velocity with which the maximum amplitude of the group advances in the medium. This is called the 'group velocity*. *Thus the group velocity is the velocity with which the energy in the group is transmitted.* The individual waves travel "inside" the group with their phase velocities.

In Fig. below are shown two plane harmonic waves of equal amplitudes but slightly different frequencies, travelling from left to right. The dotted curve represents the wave of lower frequency and is travelling faster. At a certain instant the two waves are in phase at the point *P*. Therefore, at this instant the maximum amplitude of the group formed by them also lies at *P*. At a later instant the maximum will be build up a little to the left of *P*. That is, the maximum will move to the left with time relative to the waves. As a result, the group velocity will be lower than the wave velocities.

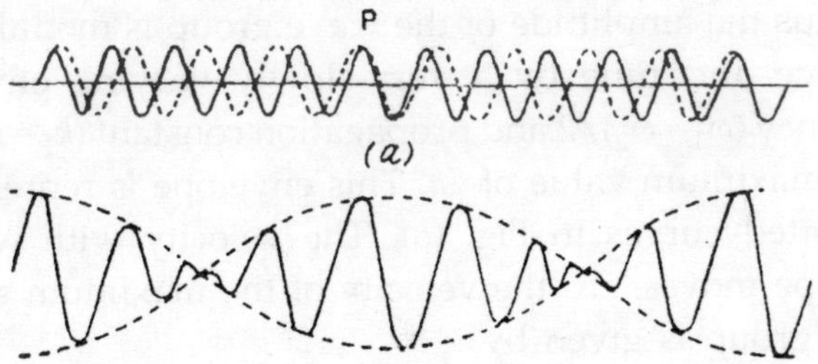

(b)

Expression for Group Velocity: Let us consider a wave group which consists of two components of equal amplitude a but slightly different angular frequencies ω_1 and ω_2, and propagation constants k_1 and k_2. Their separate displacements are given by

$$y_1 = a \sin (\omega_1 t - k_1 x), \quad \text{... (i)}$$

and

$$y_2 = a \sin (\omega_2 t - k_2 x), \quad \text{... (ii)}$$

Their superposition gives

$$y = y_1 + y_2$$
$$= a[(\sin \omega_1 t - k_1 x) + \sin(\omega_2 t - k_2 x)].$$

Using the trigonometric relation, $\sin A + \sin B = 2 \sin \frac{A+B}{2} \cos \frac{A-B}{2}$, we get

$$y = 2a \sin\left[\frac{(\omega_1+\omega_2)t}{2} - \frac{(k_1+k_2)x}{2}\right] \cos\left[\frac{(\omega_1-\omega_2)t}{2} - \frac{(k_1-k_2)x}{2}\right]$$

$$\text{or } y = 2a \cos\left[\frac{(\omega_1-\omega_2)t}{2} - \frac{(k_1-k_2)x}{2}\right] \sin\left[\frac{(\omega_1+\omega_2)t}{2} - \frac{(k_1+k_2)x}{2}\right]$$

This represents a wave system with a frequency $(\omega_1 + \omega_2)/2$ is very close to the frequency of either component but with an amplitude given by

$$A = 2a \cos\left[\frac{(\omega_1-\omega_2)t}{2} - \frac{(k_1-k_2)x}{2}\right].$$

Thus the amplitude of the wave group is modulated both in space and time by a very-slowly varying envelope of frequency $(\omega_1 - \omega_2)/2$ and propagation constant $(k_1 - k_1)/2$, and has a maximum value of $2a$. This envelope is represented by the dotted curves in Fig. (*b*). The velocity with which this envelope moves, *i.e.*, the velocity of the maximum amplitude of the group is given by

$$v_g = \frac{\omega_1 - \omega_2}{k_1 - k_2} = \frac{\Delta\omega}{\Delta k}$$

If a group contains a number of frequency-components in an infinitely small frequency-interval, then the above expression may be written as

$$\boxed{v_g = \frac{d\omega}{dk}}$$

This is the expression for the group velocity.

Relation between Group Velocity and Wave Velocity: Since $\omega = kv_p$, where v_p is wave (phase) velocity, the group velocity is given by

$$vg = \frac{d\omega}{dk} = \frac{d}{dk}(kv_p)$$

$$= v_p + k\frac{dv_p}{dk}.$$

Now, $k = \dfrac{2\pi}{\lambda}$ where λ is wavelength. Therefore

$$v_g = v_p + \frac{2\pi}{\lambda}\frac{dv_p}{d\left(\frac{2\pi}{\lambda}\right)}$$

$$= v_p + \frac{1}{\lambda}\frac{dv_p}{d\left(\frac{1}{\lambda}\right)}$$

But $d\left(\dfrac{1}{\lambda}\right) = -\dfrac{1}{\lambda^2}d\lambda$. Therefore

$$\boxed{v_g = v_p - \lambda\frac{dv_p}{d\lambda}}$$

This is the relation between group velocity v_g and wave (phase) velocity v_p in a dispersive medium.

Usually $dv_g/d\lambda$ is positive (normal dispersion). Therefore, the group velocity is smaller than the wave velocity. In case of anomalous dispersion, however, v_g is greater than v_p.

In a non-dispersive medium ($v_p = \omega/k$ = constant), we have $\dfrac{dv_p}{d\lambda} = 0$, so that

$$v_g = v_p$$

For instance, for light waves in free space the group velocity is same as the wave velocity.

Phase Velocity (or Wave Velocity) of de Broglie Waves: According to the de Broglie concept of matter waves, a particle of mass m moving with velocity v has a wave associated with it whose wavelength is given by

$$\lambda = \frac{h}{mv},$$

where h is Planck's constant. The propagation constant of the wave is

$$k = \frac{2\pi}{\lambda} = \frac{2\pi mv}{h}.$$

If E be the energy of the particle, then the frequency v of the associated wave can be specified by the quantum condition $E = hv$, so that

$$v = \frac{E}{h}$$

and the angular frequency is

$$\omega = 2\pi v = \frac{2\pi E}{h}.$$

The energy E is related to the mass m of the particle by the Einstein's relation $E = mc^2$, so that

$$\omega = \frac{2\pi mc^2}{h}$$

By definition, the de-Broglie wave velocity is given by

$$v_p = \frac{\omega}{k}$$

$$= \frac{2\pi mc^2 / h}{2\pi mv / h} = \frac{c^2}{v}.$$

But the particle velocity v is always less than c (the velocity of light). Therefore, the de-Broglie wave velocity v_p must be greater than c. This is indeed an unexpected result ! Furthermore, according to this result, the de Broglie wave-train associated with the particle would travel much faster than the particle itself and would leave the particle far behind. This is nothing but the collapse of the wave description of the particle.

Group Velocity of de Broglie Waves: The phase velocity of a wave associated with a particle comes out to be greater than the velocity of light. This difficulty is resolved by postulating that the moving particle is associated with a 'wave packet' or 'wave group', rather than a single wave-train. A wave group corresponding to a wavelength A is composed of a number of component waves with slightly different wavelengths in the neighbourhood of 3λ. The velocity of a wave group (group velocity) may be much less than the velocity of the individual components (wave velocity or phase velocity).

Now, the angular frequency and the propagation constant of de Broglie waves associated with a particle of rest mass m_0 moving with velocity v are given by

$$\omega = \frac{2\pi mc^2}{h} \text{ and } k = \frac{2\pi mv}{h},$$

where, by the theory of relativity, $m = \dfrac{m_0}{\sqrt{\left(1 - \dfrac{v^2}{c^2}\right)}}$. Thus

$$\omega = \frac{2\pi m_0 c^2}{h\sqrt{\left[1 - \left(v^2 / c^2\right)\right]}}.$$

and
$$k = \frac{2\pi m_0 v}{h\sqrt{\left[1 - \left(v^2 / c^2\right)\right]}}$$

By differentiation, we obtain

$$\frac{d\omega}{dv} = \frac{2\pi m_0 v}{h\left(1 - \dfrac{v^2}{c^2}\right)^{3/2}}$$

and
$$\frac{dk}{dv} = \frac{2\pi m_0}{h\left(1-\frac{v^2}{c^2}\right)^{3/2}}$$

The group velocity v_g of the de-Broglie waves associated with the particle is therefore

$$v_g = \frac{d\omega}{dk} = \frac{d\omega / dv}{dk / dv}.$$

Substituting the values of $d\omega/dv$ and dk/dv from above, we find that

$$v_g = v.$$

Hence the de-Broglie Wave 'group' Associated with a Moving Particle Travels with the same Velocity as the Particle: Thus a moving particle is equivalent to a wave group.

Representation of Equivalent Wave-forms: A single electron moving with a uniform velocity v is associated with a wave group moving with a (group) velocity equal to v, as shown in Fig. (*a*).

A *uniform* beam of electrons moving with equal velocities v is equivalent to a monochromatic wave having a wavelength equal to $\frac{h}{mv}$, as shown in Fig. below (*b*).

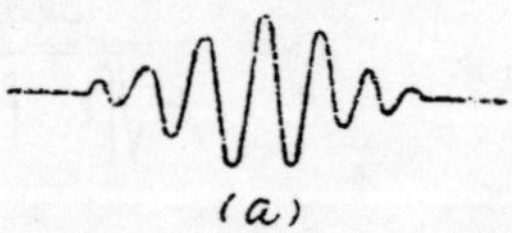

(a)

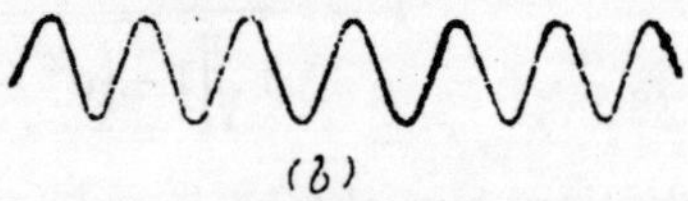

(b)

Heisenberg's Uncertainty Principle

This principle was described by Heisenberg in 1927. It states that we can measure either the position or the momentum of a particle with any desired degree of accuracy (within the limits of experimental equipment) but *it is impossible to measure both the position and momentum simultaneously with unlimited accuracy.*

Heisenberg has shown that if we design even an ideal experiment to measure simultaneously the position x and the corresponding component of momentum p_x of a particle, there will be an uncertainty of Δx in the position and an uncertainty of Δpx in the momentum such that

$$\Delta x \, \Delta px \geq \frac{h}{4\pi}$$

where h is Planck's constant. Thus the uncertainty principle states :

The product of the uncertainty Δx in the position of a body at some instant and the uncertainty Δpx in its component of momentum at the same instant is equal to or greater than $h/4\pi$. If one quantity is tried to be measured exactly, the other will become completely uncertain, and vice versa. Let us illustrate the principle by simple experiments :

Determination of the Position of a Particle by a Microscope: Suppose we imagine to determine accurately the position and momentum of an electron e (Fig.) along the x-axis, using an ideal microscope free from all mechanical and optical defects. If the particle were completely isolated from external influences, then two determinations of its exact positions at exactly measured times would give its exact velocity (and hence momentum). But a completely isolated particle cannot even be detected. We must at least illuminate it. Now, the limit of resolution of the microscope is given by

$$\Delta x = \frac{\lambda}{2 \sin i}, \quad \text{... } (i)$$

where λ is the wavelength of light illuminating the particle and *i* the semi-vertex angle of the cone of rays entering the objective lens of the microscope. Δx represents the order of the uncertainty in the *x*-component of the position of the electron as determined by the microscope.

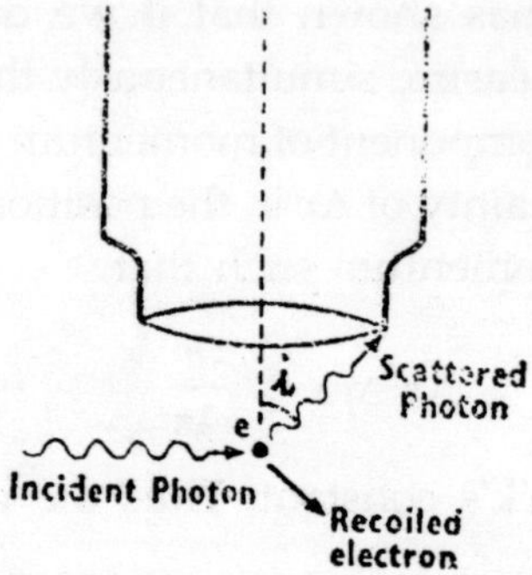

The very act of illumination, however, disturbs the electron because the light photons illuminating it transfer momentum to it. The electron therefore recoils (Compton effect) with changed momentum and any subsequent observation of the path of the electron cannot tell us precisely what was its momentum before it was illuminated. Thus uncertainty is introduced in the measurement of momentum.

Now, the minimum amount of light incident on the electron is one photon. The electron will be seen only when it scatters this photon into the microscope. But the scattered photon can enter the microscope anywhere between the angular range + *i* to – *i* (see Fig.). The momentum p of the scattered photon is, by de-Broglie relation ($\lambda = h/p$), given by $p = h/\lambda$. Hence its *x*-component can lie anywhere between $(+\ h/\lambda)\sin i$ and $(-h/\lambda)\sin i$, that is

$$\Delta p_{x_0} = \frac{2h}{\lambda}\sin i\,. \qquad \text{... }(ii)$$

The *x*-component of the momentum of the recoiling electron has the same uncertainty Δpx as for the photon since momentum is conserved in the collision. Therefore, the product of the uncertainties in the x-components of position and momentum for the electron is, from (*i*) and (*ii*),

$$\Delta x \Delta p_x = \left(\frac{\lambda}{2 \sin i}\right)\left(\frac{2h}{\lambda} \sin i\right) = h > \frac{h}{4\pi}$$

This is in agreement with the uncertainty relation. When the position of the electron is more accurately determined by reducing λ sufficiently (say, by using γ-rays), the uncertainty in the momentum of the electron becomes larger. This imaginary experiment emphasizes that the uncertainty enters into the picture through the act of observation, and not on account of any crudity in the measuring instrument.

Diffraction of Electron-beam at a Narrow Slit: Let a beam of electrons moving along the x-axis pass through a slit, and behaving as wave, produce a diffraction pattern on a screen (Fig.). In Fraunhofer's diffraction at a single slit, the direction of the first minimum is given by

$$\sin \theta_1 = \frac{\lambda}{\Delta y}, \qquad \ldots (i)$$

where Δy is the width of the slit. Electrons are diffracted at angles greater than θ_1 also, but most of them lie within the angular range $+\theta_1$ to $-\theta_1$.

The fact that a diffraction pattern appears indicate that each electron has passed through the slit, but at what place in the slit is indefinite. Thus the y coordinate of the position of the electron has an uncertainty Δy.

Initially the electrons moving along the x-axis have zero momentum along the y-direction. But, at the slit, they deviate from their initial path to form the pattern and have component of momentum in the y-direction also. The momentum of the electron is, by de-Broglie relation, $p = h/\lambda$. Hence its y-component can lie anywhere between $(+h/\lambda) \sin \theta_1$ and $(-h/\lambda) \sin \theta_1$. That is, the uncertainty in the y-component of the momentum is

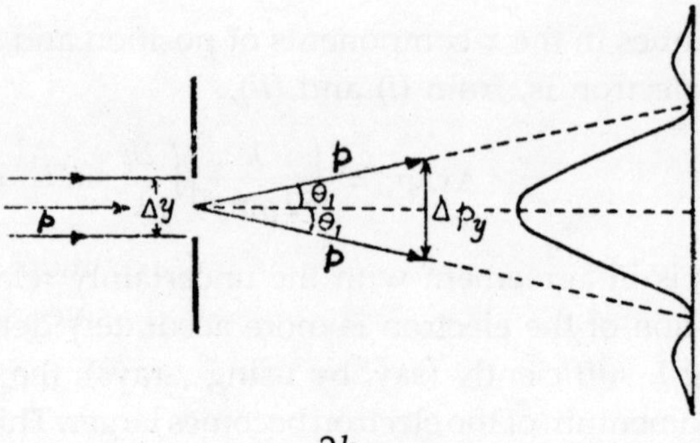

$$\Delta p_y = \frac{2h}{\lambda} \sin \theta_1$$

Substituting the value of $\sin \theta_1$ from eq. (*i*), we get

$$\Delta p_y = \frac{2h}{\lambda} \frac{\lambda}{\Delta y}$$

or $$\Delta y \, \Delta pv = 2h > \frac{h}{4\pi},$$

which is in agreement with the uncertainty principle. If we try to improve the accuracy of the position by decreasing the width of the slit, the diffraction pattern will be widened. This means that the uncertainty in momentum will increase.

The uncertainty principle is applicable to all material particles, from electrons to large bodies occurring in mechanics. In case of large bodies, however, the uncertainties are negligibly small compared to the ordinary experimental errors.

Uncertainty Principle : Heisenberg's uncertainty principle tells that it is impossible to measure both the position and the momentum of a particle simultaneously with unlimited accuracy. Heisenberg deduced that the product of the uncertainty in position, Δx, with the uncertainty in linear momentum, Δp, is of the order of the Planck's constant h. That is

$$\Delta x \, \Delta p \geq \frac{h}{4\pi}.$$

This equation is one form of a general principle put forward by Heisenberg in 1927 when electron diffraction was being investigated for the first time. The principle applies to all canonically conjugate pairs of quantities in which measurement of one quantity affects the capacity to measure the other. For example, angular position ϕ and angular momentum L form such a pair. If r is the length of the radius vector of a particle at any instant with reference to an origin, we have

$$x = r\phi$$

so that $$\Delta x = r\Delta\phi.$$

Similarly $$v = r\omega \quad \text{where } \omega \text{ is angular velocity}$$

or $$mv = mr\omega$$

or $$p = mv = \frac{L}{r} \qquad [\because L = mr^2\omega]$$

so that $$\Delta p = \frac{\Delta L}{r}.$$

$$\therefore \qquad \Delta x \Delta p = (r\Delta\phi)\left(\frac{\Delta L}{r}\right) = \Delta\phi\Delta L.$$

Since $\Delta x \Delta p \geq \frac{h}{4\pi}$; we can write

$$\Delta\phi\Delta L \geq \frac{h}{4\pi}.$$

Thus if angular momentum of an object has an uncertainty ΔL, the other quantity which is uncertain is the angular position ϕ which is uncertain by an amount $\Delta\,\phi \geq \dfrac{h}{4\pi\,\Delta L}$.

According to Bohr's orbit concept, an electron in an atom has sharply defined angular momentum, with no uncertainty whatever. This means that $\Delta L = 0$. Then according to uncertainty relation, $\Delta\,\phi$ must be infinite, that is, the angular position will be quite uncertain and in the atom the electron cannot at all be located.

Concept of Bohr Orbits and the Principle of Uncertainty: The uncertainty relation $\Delta x \, \Delta p \geq h/4\pi$ can also be written in terms of the conjugate pair of quantities, energy E and time t. We know that the kinetic energy of a particle is

$$E = \frac{1}{2} mv^2 = \frac{p^2}{2m},$$

where p is the linear momentum ($p = mv$). Thus

$$\Delta E = \frac{p \, \Delta p}{m}.$$

But $p = mv = m \dfrac{\Delta x}{\Delta t}$.

$$\therefore \qquad \Delta E = \frac{\Delta x}{\Delta t} \Delta p$$

$$\text{or} \qquad \Delta E \Delta t = \Delta x \Delta p.$$

Since $\Delta x \, \Delta p > h/4\pi$, we can write

$$\Delta E \Delta t \geq h/4\pi.$$

According to the Bohr's orbit concept, an electron in an atom revolves in one of the quantised orbits, each having a sharply defined energy, with no uncertainty whatever. That is $\Delta E = 0$. Then according to uncertainty relation, Δt must be infinite which means that all energy states of the atom must have an infinite life-time. But we know that the excited states of the atom from which radiation occurs do not last indefinitely and the life time is of the order of 10^{-8} sec. Thus the concept of Bohr orbits violates the uncertainty principle.

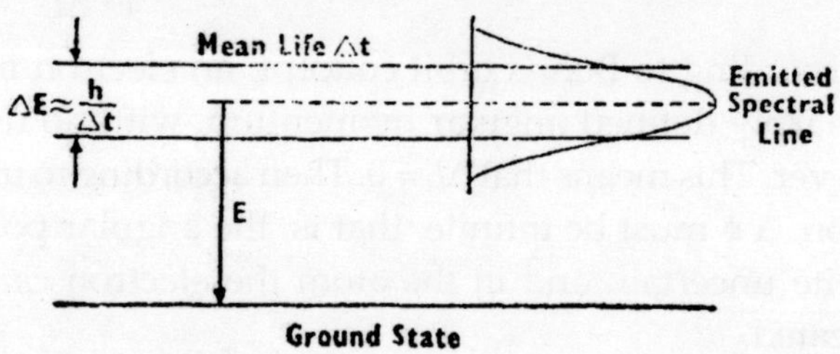

The finite life-time Δt gives a finite width (uncertainty) to the energy levels, $\Delta E \approx \dfrac{h}{4\pi\Delta t}$ (Fig.). This means that excited energy levels have a finite energy-spread so that the radiation emitted when atoms fall back to the ground state from a particular level is not truely monochromatic. In other words, the spectral lines can never be infinitely sharp but must have a natural finite width (natural broadening).

Uncertainty Principle with Macroscopic Objects: The uncertainty principle does not reveal itself when dealing with macroscopic objects. This is due to the extremely small value of h (= 6.63×10^{-34} joule-sec). In case of macroscopic objects the experimental errors involved in measuring position and momentum simultaneously, Δx and Δp, are always such that $\Delta x\ \Delta p >> h/4\pi$. Hence the fundamental limit $\Delta x\ \Delta p = h/4\pi$ is never reached due to our inability of measurement to this accuracy.

Let us consider a 1-gm bullet moving at 10^5 cm/sec. Its position at any instant might be known within an experimental accuracy of $\Delta x = \pm\ 10^{-3}$ cm and its momentum to within $\Delta p = \pm$ 10 gm-cm/sec. Then

$$\Delta x\ \Delta p = 10^{-2} \text{ erg-sec} = 10^{-9} \text{ joule-sec},$$

which is very much greater than $h/4\pi = 0.525 \times 10^{-34}$ joule-sec.

Uncertainty in Velocity: From the uncertainty principle, we have

$$\Delta x \Delta p \approx h$$

or $$\Delta p \approx \frac{h}{\Delta x}.$$

But it is given that $\Delta x = \dfrac{h}{p}$ (de-Broglie wavelength).

$\therefore$ $$\Delta p \approx p$$

or $$\Delta(mv) \approx mv$$

or $$m\Delta v \approx mv$$

$\therefore$ $$\Delta v \approx v.$$

This is what we had to show.

Electrons in Nuclei: Let us suppose that an electron is present in an atomic nucleus whose radius is 5×10^{-15} meter. For an electron to be confined within such a nucleus, the uncertainty Δx in its position is of the ordar of 5×10^{-15} meter. According to uncertainty relation ($\Delta x\ \Delta p \geq h/4\pi$), the corresponding uncertainty in the electron's momentum is

$$\Delta p \geq \frac{h}{4\pi\ \Delta x}$$

$$\geq \frac{6.63 \times 10^{-34}\ \text{joule - sec}}{4 \times 3.14 \times \left(5 \times 10^{-15}\ \text{meter}\right)}$$

$$\geq 1.1 \times 10^{-20}\ \text{kg-m/sec.}$$

If this is the uncertainty in the electron's momentum, the momentum itself must be comparable to it ($p \approx 1.1 \times 10^{-20}$ kg-meter/sec). An electron having this momentum must have a velocity comparable to the velocity of light. Hence let us apply the relativistic formula for calculating the energy of the electron. The total (kinetic + rest mass) energy of an electron with a rest mass m_0 and momentum p is given by

$$E \geq \sqrt{\left(p^2c^2 + m_0^2\ c^4\right)}$$

Now, $$pc = 1.1 \times 10^{-20}\ \text{kg-m/sec} \times (3 \times 10^8\ \text{m/sec})$$

$$= 3.3 \times 10^{-12}\ \text{joule}$$

$$= \frac{3.3 \times 10^{-12}}{1.6 \times 10^{-19}} \simeq 20 \times 10^6\ \text{eV} = 20\ \text{MeV}.$$

The rest mass energy of the electron, m_0c^2, is 0.51 MeV, which is negligible compared to the above value of pc.

Thus the electron must have a kinetic energy of the order of 20 MeV if it is to be a nuclear constituent. But since the energy of a β particle (electron) emitted from a nucleus has not been found to exceed 2-3 MeV, we conclude that electrons *cannot be* present within nuclei.

PROBLEMS

1. An electron and a photon each have a wavelength of 2 0Å. Compute their (a) momenta, (b) total energies, (c) ratio of kinetic energies, (h = 6.63 × 10^{-34} joule-sec, electron rest mass = 9.1 × 10^{-31} kg, c = 3.0 × 10^{8} meter/sec, 1 eV = 1.6 × 10^{-19} joule).

Solution : By the de Broglie wavelength formula ($\lambda = h/p$), the momentum p of the 2.0 Å-electron is

$$p = \frac{h}{\lambda} = \frac{6.63 \times 10^{-34} \text{ joule - sec}}{2.0 \times 10^{-10} \text{ meter}}$$

$$= 3.31 \times 10^{-24} \text{ kg-meter/sec.}$$

The momentum of a photon of frequency v is given by

$$p = \frac{hv}{c} = \frac{h}{\lambda}.$$

Since the wavelength λ of photon is same (2.0Å) as for electron, its momentum is also 3.31 × 10^{-24} kg-meter/sec.

(*b*) The kinetic energy of the electron is

$$K_e = \frac{p^2}{2m_0}$$

$$= \frac{\left(3.31 \times 10^{-24} \text{ kg - m / sec}\right)^2}{2 \times \left(9.1 \times 10^{-31} \text{ kg}\right)}$$

$$= 0.602 \times 10^{-17} \text{ joule}$$

$$= \frac{0.602 \times 10^{-17}}{1.6 \times 10^{-19}} = 37.6 \text{ eV.}$$

The rest energy of the electron is

$$m_0c^2 = (9.1 \times 10^{-31} \text{ kg}) \times (3.0 \times 10^8 \text{ meter/sec})^2$$

$$= 81.9 \times 10^{-15} \text{ joule}$$

$$= \frac{81.9 \times 10^{-15}}{1.6 \times 10^{-19}}$$

$$= 0.51 \times 10^6 \text{ eV} = 0.51 \text{ MeV}.$$

The total energy of the electron is the sum of the kinetic and the rest energy. Since the kinetic energy (37.6 eV) is very small compared with the rest energy, the total energy is almost the same as the rest energy 0.51 MeV.

The total energy of the photon is

$$E = hv = \frac{hc}{\lambda} = pc$$

$$= (3.31 \times 10^{-24} \text{ kg-m/s}) \times (3.0 \times 10^8 \text{ m/s})$$

$$= 9.93 \times 10^{-16} \text{ joule}$$

$$= \frac{9.93 \times 10^{-16}}{1.6 \times 10^{-19}}$$

$$= 6.21 \times 10^3 \text{ eV} = 6.21 \text{ keV}.$$

Since the rest energy of a photon is zero, the total energy is, in fact, the kinetic energy of the photon.

(*c*) The kinetic energy of the electron is

$$K_e = 37.6 \text{ eV}.$$

The kinetic energy of the photon is same as its total energy, so that,

$$K_p = 6.21 \times 10^3 \text{ eV}.$$

$$\therefore \quad \frac{K_e}{K_p} = \frac{37.6}{6.21 \times 10^3} = 6.05 \times 10^{-3}.$$

2. ***Let's find out what will be the kinetic energy of an electron if its de Broglie wavelength equals the wavelength of the yellow line of sodium (5896Å). The rest mass of electron is $m_0 = 9.1 \times 10^{-31}$ kg and $h = 6.63 \times 10^{-34}$ J-s.***

Solution : The de-Broglie wavelength of a particle of mass m moving with velocity v is given by

$$\lambda = \frac{h}{mv}.$$

Ignoring relativistic mass variation ($m = m_0$) : ... (*i*)

$$\lambda = \frac{h}{m_0 v}.$$

If K be the kinetic energy, then

$$K = \frac{1}{2} m_0 v^2.$$

Substituting the value of v in terms of λ from eq. (*i*), we get

$$K = \frac{h^2}{2m_0\lambda^2}$$

$$= \frac{\left(6.63 \times 10^{-34} \text{ joule - sec}\right)^2}{2\left(9.1 \times 10^{-31} \text{kg}\right)\left(5896 \times 10^{-10} \text{m}\right)^2}$$

$$= 6.95 \times 10^{-25} \text{ joule}$$

$$= \frac{6.95 \times 10^{-25}}{1.6 \times 10^{-19}} = 4.34 \times 10^{-6} \text{ eV}.$$

3. ***We can determine the velocity and kinetic energy of a neutron having de Broglie wavelength 1Å (mass of neutron is 1.67×10^{-27} kg, $h = 6.63 \times 10^{-34}$ joule-sec).***

Solution : The de-Broglie wavelength of a particle of mass m moving with velocity v is given by

$$\lambda = \frac{h}{mv}.$$

In the non-relativistic case $m \simeq m_0$ (rest mass). Thus

$$\lambda = \frac{h}{m_0 v}$$

or
$$v = \frac{h}{m_0 \lambda}.$$

For $m_0 = 1.67 \times 10^{-27}$ kg and $\lambda = 1\text{Å} = 10^{-10}$ m, we have

$$v = \frac{6.63 \times 10^{-34} \text{ joule-sec}}{(1.67 \times 10^{-27} \text{ kg}) \times 10^{-10} \text{ m}}$$

$$= 3.97 \times 10^3 \text{ m/sec}.$$

The kinetic energy of the particle is

$$K = \frac{1}{2} m_0 v^2$$

$$= \frac{1}{2}(1.67 \times 10^{-27} \text{ kg}) \times (3.97 \times 10^3 \text{ m/sec})^2$$

$$= 1.32 \times 10^{-20} \text{ joule}$$

$$= \frac{1.32 \times 10^{-20}}{1.6 \times 10^{-19}} = 0.0825 \text{ eV}.$$

4. We may find the de-Broglie wavelength of (a) a ball of mass 1.0 kg moving with a speed of 1.0 meter/sec. (b) an electron travelling with speed of 10^6 meter/sec.

($h = 6.63 \times 10^{-34}$ joule-sec., electron mass = 9.1×10^{-31} kg.)

Solution : The de Broglie wavelength of a particle of mass m moving with velocity v is

$$\lambda = \frac{h}{mv}.$$

If the velocity is much smaller compared with the velocity of light (non-relativistic case) then $m \simeq m_0$, so that

$$\lambda = \frac{h}{m_0 v}$$

(*a*) For ball, $m_0 = 1.0$ kg and $v = 1.0$ meter/sec.

$$\therefore \quad \lambda = \frac{6.63 \times 10^{-34} \text{ joule - sec}}{1.0 \text{ kg} \times 1.0 \text{ meter / sec}}$$

$$= 6.63 \times 10^{-34} \text{ meter} = 6.63 \times 10^{-24} \text{ Å}$$

(*b*) For electron, $m_0 = 9.1 \times 10^{-31}$ kg and $v = 10^6$ m/sec.

$$\therefore \quad \lambda = \frac{6.63 \times 10^{-34} \text{ joule - sec}}{(9.1 \times 10^{-31} \text{ kg}) \times (10^6 \text{ m / sec})}$$

$$= 7.28 \times 10^{-10} \text{ meter} = 7.28 \text{ Å}.$$

5. *Let's calculate the de-Broglie wavelength of an electron whose kinetic energy is 50 eV. h = 6.62 × 10⁻³⁴ joule-sec, m₀ = 9.1 × 10⁻³¹ kg, 1 eV = 1.6 × 10⁻¹⁹ joule.*

Solution : The de-Broglie wavelength of a particle moving with velocity v is given by

$$\lambda = \frac{h}{mv},$$

where m is the relativistic mass of the particle.

The given energy of 50 eV is small compared with the rest energy (0.51 MeV) of electron, so that we can ignore relativistic effect and can put $m \approx m_0$. Then

$$\lambda = \frac{h}{m_0 v}. \qquad \text{... } (i)$$

If K be the kinetic energy of the particle, then

$$K = \frac{1}{2} m_0 v^2$$

or $$v = \sqrt{\left(\frac{2K}{m_0}\right)}.$$

Substituting this value of v in eq. (*i*), we get

$$\boxed{\lambda = \frac{h}{\sqrt{(2m_0 K)}}.}$$

This is the expression for the de-Broglie wavelength of a particle having kinetic energy K, *in the non-relativistic case.*

Now, $K = 50$ eV $= 50 \times 1.6 \times 10^{-19}$ joule and $m_0 = 9.1 \times 10^{-31}$ kg. Therefore

$$\lambda = \frac{6.62 \times 10^{-34} \text{ joule - sec}}{\sqrt{\left[2 \times \left(9.1 \times 10^{-31} \text{ kg}\right) \times \left(50 \times 1.6 \times 10^{-19} \text{ joule}\right)\right]}}$$

$$= \frac{6.62 \times 10^{-34} \text{ joule - sec}}{3.82 \times 10^{-24} \text{ kg - m / sec}}$$

$$= 1.73 \times 10^{-10} \text{ meter} = 1.73 \text{ Å}.$$

Alternative Method

The de-Broglie wavelength of a V-volt electron is given by (ignoring relativistic effects)

$$\lambda = \frac{12.27}{\sqrt{V}} = \sqrt{\left(\frac{150}{V}\right)} \text{ Å}.$$

Here the given electron is 50-volt electron.

$$\therefore \quad \lambda = \sqrt{\left(\frac{150}{50}\right)} = \sqrt{3} = 1.73 \text{ Å}$$

6. Each of a photon and an electron has an energy of 1 keV (= 1000 eV). Their corresponding wavelengths ($h = 6.6 \times 10^{-34}$ joule-sec, $m_0 = 9.1 \times 10^{-31}$ kg, leV $= 1.6 \times 10^{-19}$ joule) may be calculated.

Solution : The energy of a photon is given by

$$E = h\nu = \frac{hc}{\lambda}$$

Its wavelength is thus

$$\lambda = \frac{hc}{E}.$$

Putting the given values :

$$\lambda = \frac{\left(6.6 \times 10^{-34} \text{ joule - sec}\right)\left(3.0 \times 10^{8} \text{ meter / sec}\right)}{\left(1000 \times 1.6 \times 10^{-19} \text{ joule}\right)}$$

$$= 12.4 \times 10^{-10} \text{ meter} = 12.4 \text{ Å}.$$

If K be the non-relativistic kinetic energy of the electron, then its de-Broglie wavelength is given by

$$\lambda = \frac{h}{\sqrt{(2m_0K)}}.$$

Putting the given values:

$$\lambda = \frac{6.6 \times 10^{-34} \text{ joule - sec}}{\sqrt{\left[2 \times \left(9.1 \times 10^{-31} \text{ kg}\right) \times \left(1000 \times 1.6 \times 10^{-19} \text{ joule}\right)\right]}}$$

$$= \frac{6.6 \times 10^{-34} \text{ joule - sec}}{17 \times 10^{-24} \text{ kg - m / sec}} = 0.39 \times 10^{-10} \text{ m} = 0.39 \text{ Å}$$

7. Let's calculate the de-Broglie wavelength of an electron accelerated through a potential difference of (i) 100 volts, (ii) 300 volts, (iii) 1 kV.

Solution : The de-Broglie wavelength of an electron accelerated through V volt in non-relativistic motion is given by

$$\lambda = \sqrt{\left(\frac{150}{V}\right)} \text{ Å}$$

(*i*) $$\lambda = \sqrt{\left(\frac{150}{100}\right)} = 1.22 \text{ Å}.$$

(*ii*) $$\lambda = \sqrt{\left(\frac{150}{300}\right)} = 0.71.$$

(*iii*) $$\lambda = \sqrt{\left(\frac{150}{1000}\right)} = 0.39 \text{ Å}.$$

8. An electron microscope is operated at 10,000 volts. The wavelength associated with the electron beam is calculated as follows considering non-relativistic motion.

Solution : In non-relativistic motion, the de-Broglie wavelength of V-volt 'electron' is given by

$$\lambda = \frac{h}{\sqrt{(2m_0eV)}} = \frac{12.27}{\sqrt{V}} \text{ Å} = \sqrt{\left(\frac{150}{V}\right)} \text{ Å}$$

$$= \sqrt{\left(\frac{150}{10000}\right)} = 0.122 \text{ Å}$$

9. X-rays of wavelength 0.82 Å fall on a metal plate. The wavelength associated with the photoelectrons emitted may be found out here. We can neglect the work function (h = 6.6 × 10^{-34} J-s, c = 3.0 × 10^8 m/s, m_0 = 9.1 × 10^{-31} kg).

Solution : In the absence of work function, the kinetic energy K of a photoelectron is the same as the energy of an incident X-ray photon, that is,

$$K = h\upsilon = \frac{hc}{\lambda}$$

$$= \frac{(6.6 \times 10^{-34} \text{ joule-sec})(3.0 \times 10^8 \text{ meter-sec}^{-1})}{0.82 \times 10^{-19} \text{ meter}}$$

$$= 2.4 \times 10^{-15} \text{ joule.}$$

The de-Broglie wavelength of electron of energy K is given by

$$\lambda = \frac{h}{\sqrt{(2m_0K)}}$$

$$= \frac{6.6 \times 10^{-34} \text{ joule-sec}}{\sqrt{\left[2 \times \left(9.1 \times 10^{-31} \text{ kg}\right) \times \left(2.4 \times 10^{-15} \text{ joule}\right)\right]}}$$

$$= 1 \times 10^{-11} \text{ meter} = 0.1 \text{ Å}.$$

10. Let's calculate the glancing angle at which electrons of energy 100 eV must fall on the lattice planes of a crystal to obtain a strong Bragg reflection in the first order. The lattice spacing is 2.15 Å.

The wavelength of the 100-volt electrons is given by

$$\lambda = \sqrt{\left(\frac{150}{100}\right)} \text{ Å} = 1.22 \text{ Å}.$$

The Bragg equation is

$$2d \sin \theta = n\lambda,$$

where d is lattice spacing. For the first order, $n = 1$. Thus

$$\sin \theta = \frac{\lambda}{2d} = \frac{1.22 \text{ Å}}{2 \times 2.15 \text{ Å}} = 0.284$$

$$\therefore \qquad \theta = \sin^{-1}(0.284) = 16.5°$$

11. The de-Broglie wavelength of a neutron of energy 12.8 MeV. (mass of neutron = 1.675 × 10^{-27} kg. h = 6.62 × 10^{-34} joule-sec. c = 3.0 × 10^{8} cm/sec and 1 eV = 1.6 × 10^{-13} joule.) may found out.

The rest mass of the neutron is 1.675 × 10^{-29} kg, so that its rest energy is

$$m_0c^2 = (1.675 \times 10^{-27} \text{ kg}) \times (3.0 \times 10^{8} \text{ m/sec})^2$$

$$= 15.075 \times 10^{-11} \text{ joule}$$

$$= \frac{15.075\ 10^{-11}}{1.6 \times 10^{-19}} = 9.42 \times 10^8 \text{ eV} = 942 \text{ MeV}.$$

The given energy of 12.8 MeV is very small compared with the rest mass energy and so relativistic effect can be ignored.

Now, the de-Broglie wavelength of a particle with kinetic energy K, in the non-relativistic case, is given by

$$\lambda = \frac{h}{\sqrt{(2m_0K)}}$$

Here K = 12.8 MeV = 12.8 × 10^6 eV

= 12.8 × 10^6 × 1.6 × 10^{-19} joule.

$$\therefore \quad \lambda = \frac{6.62 \times 10^{-31} \text{ joule - sec}}{\sqrt{\left[2 \times \left(1.675 \times 10^{-27} \text{ kg}\right) \times \left(12.8 \times 10^6 \times 1.6 \times 10^{-19} \text{ joule}\right)\right]}}$$

$$= \frac{6.62 \times 10^{-34} \text{ joule - sec}}{8.28 \times 10^{-20} \text{ kg - m / sec}}$$

$$= 0.8 \times 10^{-4} \text{ m} = 0.8 \times 10^{-10} \text{ Å}.$$

12. The de Broglie wavelength of a 3.6 MeV-alpha particle may be found out. The rest mass of the α-particle is 6.65 × 10^{-27} kg. (h = 6.62 × 10^{-34} joule-sec, c = 3.0 × 10^8 meter/ sec and 1 eV = 16 × 10^{-10} joule.)

Solution : The given α-particle energy 3.6 MeV is very small compared with its rest mass energy and so relativistic effect can be ignored. The de-Broglie wavelength is now given by

$$\lambda = \frac{h}{\sqrt{(2m_0K)}}$$

where m_0 is the rest mass and K the kinetic energy of the particle. Here m_0 = 6.65 × 10^{-27} kg and K = 3.6 MeV = 3.6 × 1.6 × 10^{-13} joule.

$$\therefore \quad \lambda = \frac{6.62 \times 10^{-34} \text{ joule - sec}}{\sqrt{\left[2 \times \left(6.65 \times 10^{-27} \text{ kg}\right) \times \left(3.6 \times 1.6 \times 10^{-13} \text{ joule}\right)\right]}}$$

$$= \frac{6.62 \times 10^{-34} \text{ joule - sec}}{8.75 \times 10^{-20} \text{ kg - m / sec}}$$

$$= 7.56 \times 10^{-15} \text{ meter} = 7.56 \times 10^{-5} \text{ Å}.$$

13. ***Let's calculate the wavelength associated with a (i) 1 MeV (kinetic energy) electron, (ii) 1 MeV (kinetic energy) proton and (iii) 1 MeV photon. The rest masses of electron and proton are 9.1 × 10^{-31} kg and 1.67 × 10^{-27} kg respectively. (h = 6.62 × 10^{-34} joule-sec, c = 3.0 × 10^{8} meter/sec, 1 eV = 1.6 × 10^{-19} joule).***

Solution : (*i*) The rest mass energy of the electron is

$$m_0c^2 = (9.1 \times 10^{-31} \text{ kg}) \times (3.0 \times 10^8 \text{ m/sec})^2$$

$$= 81.9 \times 10^{-15} \text{ joule}$$

$$= \frac{81.9 \times 10^{-15}}{1.6 \times 10^{-19}} = 0.51 \times 10^6 \text{ eV} = 0.51 \text{ MeV}.$$

The given kinetic energy K = 1 MeV is comparable with the rest energy and so we must take the relativistic variation of mass into account. Now we shall have to use the basic formula $\lambda = h/mv$, where m is the "relativistic mass".

Let us first calculate the (relativistic) mass m of the electron of kinetic energy 1 MeV. The total energy of this electron is

$$E = K + m_0c^2$$

$$= 1 + 0.51 = 1.51 \text{ MeV} = 1.51 \times 10^6 \text{ eV}.$$

But $E = mc^2$.

$$\therefore \quad m = \frac{E}{c^2} = \frac{1.51 \times 10^6 \times 1.6 \times 10^{-19} \text{ joule}}{\left(3.0 \times 10^8 \text{ meter / sec}\right)^2}$$

$$= 2.68 \times 10^{-30} \text{ kg}.$$

Next, we base to find the velocity v of this electron. We use the formula

$$m = \frac{m_0}{\sqrt{\left(1 - \frac{v^2}{c^2}\right)}}$$

so that

$$1 - \frac{v^2}{c^2} = \frac{(m_0)^2}{m^2} = \frac{\left(9.1 \times 10^{-31}\right)^2}{\left(2.68 \times 10^{-30}\right)^2} = 0.1153$$

$$\text{or} \quad \frac{v^2}{c^2} = 1 - 0.1153 = 0.8847$$

$$\text{or} \quad \frac{v}{c} = \sqrt{(0.8847)} = 0.9406$$

$$\therefore \quad v = 0.9406 \times (3 \times 10^8)$$

$$= 2.82 \times 10^8 \text{ meter/sec.}$$

The de-Broglie wavelength of the 1-MeV electron is thus

$$\lambda = \frac{h}{mv}$$

$$= \frac{6.62 \times 10^{-34} \text{ joule - sec}}{\left(2.68 \times 10^{-30} \text{ kg}\right) \times \left(2.82 \times 10^8 \text{ meter / sec}\right)}$$

$$= 8.76 \times 10^{-13} \text{ meter}$$

$$= 8.76 \times 10^{-3} \text{ Å}.$$

(*ii*) The rest energy of the proton is

$$m_0c^2 = (1.67 \times 10^{-27} \text{ kg}) \times (3.0 \times 10^8 \text{ meter/sec})^2$$

$$= 1.5 \times 10^{-10} \text{ joule}$$

$$= \frac{1.5 \times 10^{-10}}{1.6 \times 10^{-19}} \approx 937 \times 10^6 \text{ eV} = 937 \text{ MeV}.$$

The given kinetic energy 1 MeV of the proton is much less than its rest energy so that in this case the relativistic effect can be ignored. We can therefore use the non-relativistic formula

$$\lambda = \frac{h}{\sqrt{(2m_0 K)}}.$$

$$\text{Thus } \lambda = \frac{6.62 \times 10^{-34} \text{ joule-sec}}{\sqrt{\left[(2 \times 1.67 \times 10^{-27} \text{kg}) \times (1 \times 10^{6} \times 1.6 \times 10^{-19} \text{joule})\right]}}$$

$$= \frac{6.62 \times 10^{-34} \text{ joule - sec}}{2.31 \times 10^{-20} \text{ kg - meter / sec}}$$

$$= 2.86 \times 10^{-14} \text{ meter} = 2.86 \times 10^{-4} \text{ Å}.$$

(*iii*) The rest mass of the photon and hence the rest energy is zero. Its energy is entirely kinetic and is given by

$$E = hv = \frac{hc}{\lambda}.$$

$$\therefore \quad \lambda = \frac{hc}{E}$$

$$= \frac{(6.62 \times 10^{-34} \text{ joule - sec}) \times (3 \times 10^{8} \text{ meter / sec})}{1 \times 10^{6} \times 1.6 \times 10^{-19} \text{ joule}}$$

$$= 1.24 \times 10^{-12} \text{ meter}$$

$$= 1.24 \times 10^{-2} \text{ Å}.$$

14. ***A particle of rest mass m_0 has a kinetic: energy K. shows that its de Broglie wavelength is given by***

$$\lambda = \frac{hc}{\sqrt{\left[K\left(K + 2m_0 c^2\right)\right]}}$$

Hence we may calculate the wavelength of an electron of kinetic energy 1 MeV and find what will be the value of λ if $K << m_0 c^2$

Solution : The de-Broglie wavelength of a particle moving with velocity v is given by

$$\lambda = \frac{h}{mv}, \quad \text{... (i)}$$

where m is the 'relativistic mass of the particle. By relativity, we have

$$m = \frac{m_0}{\sqrt{\left(1 - \frac{v^2}{c^2}\right)}}$$

or $$1 - \frac{v^2}{c^2} = \frac{m_0^2}{m^2}$$

or $$\frac{v^2}{c^2} = 1 - \frac{m_0^2}{m^2} = \frac{m^2 - m_0^2}{m^2}$$

or $$mv = c\sqrt{\left(m^2 - m_0^2\right)}.$$

Putting this in eq. (*i*), we get

$$\lambda = \frac{h}{c\sqrt{\left(m^2 - m_0^2\right)}} = \frac{hc}{c^2\sqrt{\left(m^2 - m_0^2\right)}} \quad \text{... (ii)}$$

Now,

$$c^2\sqrt{\left(m^2 - m_0^2\right)} = \sqrt{\left[c^4\left(m - m_0\right)\left(m + m_0\right)\right]}$$

$$= \sqrt{\left[\left\{(m - m_0)c^2\right\}\left\{(m - m_0)c^2 + 2m_0c^2\right\}\right]}$$

$$= \sqrt{\left[K\left(K + 2m_0c^2\right)\right]}. \quad [\because K = (m - m_0)\,c^2].$$

Substituting this value in eq, (*ii*), we get

$$\boxed{\lambda = \frac{hc}{\sqrt{\left[K\left(K + 2m_0c^2\right)\right]}}} \quad \text{... (iii)}$$

This is the required formula.

For an electron, $m_0c^2 \cong$ MeV, so that when $K = 1$ MeV, we have

$$\lambda = \frac{hc}{\sqrt{2}\text{MeV}}$$

$$= \frac{(6.62 \times 10^{-34} \text{ joule-sec}) \times (3.0 \times 10^{8} \text{ m/sec})}{1.414 \times 1.6 \times 10^{-13} \text{ joule}}$$

$$= 8.78 \times 10^{-13} \text{ meter} = 8.78 \times 10^{-3} \text{ Å}.$$

If $K << m_0c^2$, we can replace $K + 2m_0c^2$ by $2\,m_0c^2$ in the above formula. Then we get

$$\lambda = \frac{hc}{\sqrt{(2m_0Kc^2)}} = \frac{h}{\sqrt{(2m_0K)}}.$$

15. The radius of the first Bohr orbit in hydrogen atom is 0.53 × 10⁻¹⁰ m. Let us find the velocity of the electron in that orbit, using de-Broglie theory. (h = 6.6 × 10⁻³⁴ joule-sec, m = 9.1 × 10⁻³¹ kg).

Solution : The electron wave can be adjusted around an orbit only when the circumference of the orbit equals an integral multiple of the wavelength, that is,

$$2\pi r = n\lambda$$

For the first Bohr orbit, $n = 1$ and from de Broglie hypothesis, $\lambda = h/mv$, where v is the velocity in that orbit. Thus

$$2\pi r = \frac{h}{mv}$$

or
$$v = \frac{h}{2\pi rm}$$

$$= \frac{6.6 \times 10^{-34} \text{ joule-sec}}{2 \times 3.14 \times (0.53 \times 10^{-10} \text{ m}) \times (9.1 \times 10^{-31} \text{ kg})}$$

$$= 2.2 \times 10^{6} \text{ m/s}.$$

16. A microscope using photons is employed to locate an electron in an atom to within a distance of 0.2 Å. The uncertainty in the momentum of the electron located in this way may be found out and we can see what would be the uncertainty in velocity ($h = 6.63 \times 10^{-34}$ joule-sec, $m_0 = 9.1 \times 10^{-31}$kg).

Solution : The maximum uncertainty in electron's position is

$$\Delta x = 0.2 \text{ Å} = 0.2 \times 10^{-10}\text{m},$$

because the electron is somewhere within the distance of 0.2 Å. Hence, using uncertainty relation $\Delta x \, \Delta p = h/4\pi$, the corresponding (minimum) uncertainty in electron's momentum is

$$\Delta p = \frac{p}{4\pi \, \Delta x} = \frac{6.63 \times 10^{-34} \text{ joule-sec}}{4 \times 3.14 \times \left(0.2 \times 10^{-10} \text{ m}\right)}$$

$$= 2.64 \times 10^{-24} \text{ kg-meter/sec.}$$

The uncertainty in the velocity of the electron is

$$\Delta v = \frac{\Delta p}{m_0} = \frac{2.64 \times 10^{-24} \text{ kg-meter/sec}}{9.1 \times 10^{-31} \text{ kg}}$$

$$= 2.90 \times 10^{8} \text{ meter/sec.}$$

17. Comparison between the uncertainties in the velocities of an electron and a proton confined to a 1-nm box is made as follows. Their masses are 9.10 × 10^{-31} kg and 1.67 × 10^{-27} kg respectively.

Solution : Since uncertainty in position, Δx, is same for both, the uncertainty in momentum, Δp, is also same. The uncertainties in velocity are

$$(\Delta v)_e = \frac{\Delta p}{m_e} \text{ and } (\Delta v)_p = \frac{\Delta p}{m_p}$$

$$\therefore \quad \frac{(\Delta v)_e}{(\Delta v)_p} = \frac{m_p}{m_e} = \frac{1.67 \times 10^{-27} \text{ kg}}{9.10 \times 10^{-31} \text{ kg}} = 1835.$$

19. *A proton is confined to a nucleus of radius 5 × 10⁻¹⁵ meter. The minimum uncertainty in its momentum may be calculated. We can also calculate the minimum kinetic energy the proton should have. The proton mass is 1.67 × 10⁻²⁷ kg.*

Solution : For a proton to be confined within such a nucleus of radius 5×10^{-15} m, the maximum uncertainty in its position is

$$\Delta x = 5 \times 10^{-15} \text{ m}$$

Using uncertainty relation $\Delta x \Delta p = \frac{h}{4\pi}$, the corresponding (minimum) uncertainty in its momentum is, therefore,

$$\Delta p = \frac{h}{4\pi \Delta x} = \frac{6.63 \times 10^{-34} \text{ joule-sec}}{4 \times 3.14 \times (5 \times 10^{-15} \text{ m})}$$

$$= 1.05 \times 10^{-20} \text{ kg-m/sec.}$$

Since the *magnitude* of the momentum cannot be less than that of the uncertainty, the momentum p of the proton must be at least equal to Δp. Thus

$$p = 1.05 \times 10^{-20} \text{ kg-m/sec.}$$

The corresponding (minimum) kinetic energy of the proton is

$$K = \frac{p^2}{2m} = \frac{(1.05 \times 10^{-20} \text{ kg-m/sec})^2}{2 \times 1.67 \times 10^{-27} \text{ kg}}$$

$$= 3.3 \times 10^{-14} \text{ joule}$$

$$= \frac{3.3 \times 10^{-14}}{1.6 \times 10^{-19}} = 2 \times 10^5 \text{ eV} = 0.2 \text{ MeV.}$$

20. The position and momentum of a 1-keV electron are simultaneously measured. If its position is located to within 1 Å, Let us see as to what is the percentage of uncertainty in its momentum (electron mass is 9.1 × 10^{-31} kg).

Solution : The uncertainty in position is $\Delta x = 1$ Å $= 10^{-10}$ meter. By Heisenberg principle, the corresponding uncertainty in momentum is

$$\Delta p = \frac{h/4\pi}{\Delta x}$$

$$= \frac{\left(6.63\times10^{-34}\,\text{joule-sec}\right)/\left(4\times3.14\right)}{10^{-10}\,\text{meter}}$$

$$= 0.53\times10^{-24}\ \text{kg-m/sec.}$$

The given energy $K = 1$ keV $= 10^3$ eV is much less than the rest energy (0.51×10^6 eV) so that the relativistic effect may be ignored. We may now calculate the momentum of the electron from the non-relativistic formula,

$$p = \sqrt{(2m_0K)}$$

$$= \sqrt{\left[2\times\left(9.1\times10^{-31}\,\text{kg}\right)\times\left(10^3\times1.6\times10^{-19}\,\text{joule}\right)\right]}$$

$$= 1.71\times10^{-23}\ \text{kg-m/sec.}$$

Hence the percentage of uncertainty is

$$\frac{\Delta p}{p}\times100 = \frac{0.53\times10^{-24}}{1.71\times10^{-23}}\times100 = 3.1\%$$

21. An electron has a speed of 6.6 × 10^{4} m/s within accuracy of 0.01% we may calculated the uncertainty in the position of the electron. (Mass of electron = 9 × 10^{-31} kg, Planck's constant = 6.6 × 10^{-34} J-s).

Solution : The momentum of the electron is

$$p = mv = (9\times10^{-31}\ \text{kg})\,(6.6\times10^4\ \text{m/s})$$

$$= 59.4\times10^{-27}\ \text{kg-m/s.}$$

The uncertainty in p is 0.01 % of this value. That is,

$$\Delta p = \frac{0.01}{100} \times \left(59.4 \times 10^{-27} \text{ kg-m/s}\right)$$

$$= 59.4 \times 10^{-31} \text{ kg-m/s}$$

From Heisenberg uncertainty relation $\Delta x \, \Delta p \geq h/4\pi$ certainty in position is

$$\Delta x \geq \frac{h}{4\pi \, \Delta p}$$

$$= \frac{6.6 \times 10^{-34} \text{ J-s}}{4 \times 3.14 \times 59.4 \times 10^{-31} \text{ kg-m/s}}$$

$$= 8.8 \times 10^{-6} \text{ m.}$$

22. *Wavelengths can be determined with accuracies of one part in 10^8. Let' us see what is the uncertainty in the position of a 1-Å X-ray photon when its wavelength is simultaneously measured*

Solution : By uncertainty principle, we have

$$\Delta x \Delta p \geq \frac{h}{4\pi}. \qquad \text{... (i)}$$

The momentum p of a photon is related to its wavelength λ by

$$p = \frac{h}{\lambda}$$

or $$p\lambda = h.$$

If Δp and $\Delta\lambda$ are uncertainties in position and wavelength, we get by differentiation

$$p\Delta\lambda + \lambda \, \Delta p = 0$$

or $$\Delta p = -\frac{p\Delta\lambda}{\lambda}$$

$$= -\frac{p\Delta\lambda}{\lambda^2} \qquad \left[\because p = \frac{h}{\lambda}\right]$$

Replacing Δp by $h\,\Delta\lambda/\lambda^2$ in the uncertainty relation (*i*), we get

$$\boxed{\Delta x\,\Delta\lambda \geq \frac{\lambda^2}{4\pi}} \qquad \dots (ii)$$

This is the uncertainty relation for simultaneous measurement of position and wavelength.

The uncertainty in wavelength is given to be in the ratio $1/10^6$, so that uncertainty in a wavelength λ is given by

$$\frac{\Delta\lambda}{\lambda} = \frac{1}{10^6} = 10^{-6}.$$

Putting this value in eq. (*ii*), we get the corresponding uncertainty in position as

$$\Delta x = \frac{\lambda^2}{4\pi\Delta\lambda}$$

$$= \frac{\lambda^2}{4\pi\left(10^{-6}\right)}$$

Here $\lambda = 1\ \text{Å} = 10^{-10}$ meter.

$$\therefore \qquad \Delta x = \frac{10^{-10}\,\text{meter}}{4 \times 3.14 \times 10^{-6}}$$

$$= 8 \times 10^{\circ}\ \text{meter},$$

An excited atom has an average life-time of 10^{-8} second. That is, during this period it emits a photon and returns to the ground state. The minimum uncertainty in the frequency of this photon may be found out.

The uncertainty relation for the energy and time is

$$\Delta E\ \Delta t = h/4\pi.$$

The Photon's energy is $E = h\nu$. Therefore, the above relation can be written as

$$\Delta v \Delta t \geq \frac{1}{4\pi}$$

or $$\Delta v \geq \frac{1}{4\pi \, \Delta t}.$$

Here $\Delta t = 10^{-8}$ sec.

$$\therefore \quad \Delta v \geq \frac{1}{4 \times 3.14 \times \left(10^{-8}\,\text{sec}\right)}$$

$$\geq 8 \times 10^{6}\ \text{sec}^{-1}.$$

The *minimum* uncertainty in frequency is $8 \times 10^{3}\ \text{sec}^{-1}$.

23. The life-time of an excited state of a nucleus is 10^{-12} sec. The uncertainty in energy of γ-ray photon emitted may be calculated. ($h/2\pi = 1.054 \times 10^{-34}$ joule-sec, 1 eV = 1.6×10^{-19} joule).

Solution : The time-interval Δt in which a photon of energy uncertainty (ΔE is emitted is given by

$$\Delta E\ \Delta t \geq \frac{h}{4\pi}$$

or $$\Delta E \geq \frac{h/2\pi}{2\Delta t}.$$

The minimum uncertainty is

$$\Delta E = \frac{h/2\pi}{2\Delta t} = \frac{1.054 \times 10^{-34}\ \text{joule-sec}}{2 \times 10^{-12}\ \text{sec}}$$

$$= 0.527 \times 10^{-22}\ \text{joule}$$

$$= \frac{0.527 \times 10^{-22}}{1.6 \times 10^{-19}}$$

$$= 3.3 \times 10^{-4}\ \text{eV}.$$

24. ***If an excited state of hydrogen atom has a life-time of 2.5 × 10^{-14} sec, The minimum error with which the energy of this state can be measured is calculated as follows (h = 6.63 × 10^{-34} J-s).***

Solution : The energy of the excited state is not precisely measurable because only a finite time is available for measurement. That is, the atom does not stay in an excited state for an indefinite time but decays to the ground state, emitting a photon.

From the uncertainty relation $\Delta E\ \Delta T \geq h/4\pi$, with Δt equal to the mean life-time of the excited state, we have

$$\Delta E \geq \frac{h}{4\pi\ \Delta T} = \frac{6.63 \times 10^{-34}\ \text{joule - sec}}{4 \times 3.14 \times \left(2.5 \times 10^{-14}\ \text{sec}\right)}$$

$$= 0.21 \times 10^{-20}\ \text{joule}$$

$$= \frac{0.21 \times 10^{-20}}{1.6 \times 10^{-19}} = 0.013\ \text{eV}.$$

This is the *minimum* uncertainty in energy measurement.

7

Photon Theory

Perfectly Black Body

A perfectly black body is one which absorbs all the thermal radiation incident upon it and, because it does not reflect light, appears black. No existing body is perfectly black. An object coated with a diffuse layer of black pigment is nearly a black body.

Black Body in Practice : A cavity in a body, connected to the outside by small hole, is a practical black body (Fig.). Radiation outside the cavity enters it through the hole. This radiation strikes the inner wall (1) v, here it is partly absorbed and partly reflected. The reflected part strikes another part of the inner wall (2) and the process is repeated, until the incident radiation is totally absorbed by the wall. Since the area of the hole is negligible compared to the total area of the cavity wall, the part of the incoming radiation that can get out through the hole can be neglected. Thus all the radiation incident on the hole from the outside, is absorbed by it. Hence the hole behaves as a black body. At low temperature the hole appears black.

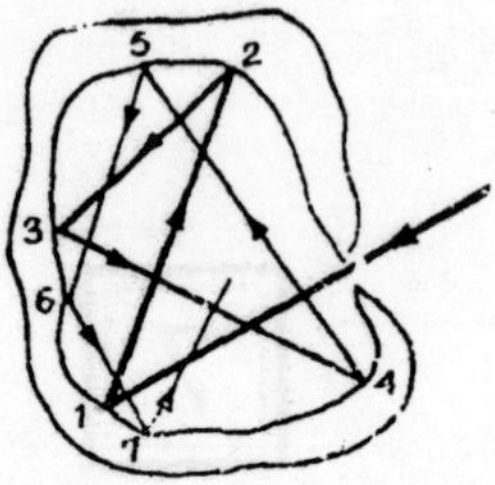

Black-Body Radiation and its Characteristics :The radiation emitted by a *heated* black body is called 'black-body radiation.' If the walls of the cavity shown in Fig. are heated *uniformly*, the hole becomes self-luminous. The inner walls emit thermal radiation into the cavity which undergoes many successive reflections and is eventually re-absorbed. In the state of thermal equilibrium the emission rate is equal to the absorption rate, and a constant amount is in transit inside the cavity. Some very small part of this radiation emerges through the hole. Since the hole behaves as a black-body, the radiation emitted by it is the black-body radiation. It is also known as cavity radiation'. It has following important characteristics :

(*i*) *The cavity radiation is more intense than the radiation from a non-black body at the same temperature* This is why the hole appears brighter than the outer wall (non-black body) of the cavity.

(*ii*) *At a given temperature, the cavity radiancy is independent of the material, shape and size of the cavity.* If this were not so, then on joining together two cavities *A* and *B* made of different materials but heated to the same temperature, a net amount of radiant energy would have flown from *A* to *B* (say). This would heat up *B* and cool down *A*. Then *B* would be used as the source and *A* as the sink of a heat-engine and work could be obtained until *A* and *B* were again at the same temperature. This process could be repeated as many times as desired, thus obtaining a continuous supply of work. This is clearly a violation of the second law of thermodynamics. Hence we conclude that the radiancy of two cavities at the same temperature but of different materials cannot be different from each other. (On the other hand, the radiancy of the outer surface does depend upon the material).

(*iii*) *The cavity radiancy E is directly proportional to the fourth power of the Kelvin temperature T of the cavity* (Stefan's law). That is

$$E = \sigma T^4$$

where σ is the universal Stefan-Boltamann constant. (On the other hand, the radiancy of the outer surface is $e\sigma T^4$, where e is the 'emissivity' which depends upon the material and upon the temperature).

(*iv*) *The spectral distribution of energy in the black body radiation at a given temperature is independent of the material, shape and size of the body.*

Cavity Hole as Black Body

A black-body is one that completely absorbs all the radiation, whatever be the wavelength, falling on it and reflects none. Now, any radiation from outside falling on the fine hole made in the walls of a cavity radiator is completely absorbed by successive reflections at the inner walls and none is reflected back to the eye. Hence, the hole behaves as a black-body.

Interior of the Cavity Radiator : The hole of a cavity radiator behaves as a black-body. On looking into it, it reflects no light into the observer's eye and appears as black. Hence the details of the interior of the cavity are not seen.

Coal Pockets Brighter : The radiancy of the hole of a cavity radiator, given by σT^4, is larger than the radiancy of the outer surface which is given by $e\sigma T^4$, where e is the surface emissivity and is less than 1. Therefore, the radiation coming out from the 'pockets' formed by the glowing coal is more intense than that coming from the surface of the coals, although both are at the same temperature. Hence the pockets appear brighter than the coal themselves.

Stefan's Law

The Stefan's fourth power law ($E = \sigma T^4$) is exact for true black bodies only. Now, only certain incandescent bodies, like platinum black, are near a true black body. Others are far away. Hence all incandescent bodies do not obey this law.

Stefan's Law for Defining Temperature : A body heated to a temperature say 50° C, cannot be treated as a black body and so the Stefan's law will not be applicable to it. Hence we cannot use this law for defining this temperature.

Spectral Distribution of Energy in Black-Body Radiation: The distribution of energy among the various wavelengths in black-body radiation was investigated by Lummer and Pringsheim in 1899. They used an electrically-heated chamber with a small aperture as the black body and measured its temperature by a thermocouple. The whole arrangement is shown in Fig.

The radiation from the black body O is made to fall on a slit *S* by means of a concave mirror .4. The slit *S* is in the focal plane of a second concave mirror *B* which reflects the radiation as a parallel beam which falls on a fluor spar prism *P*. The emerging beam falls on a third concave mirror *C* which directs it on to a Lummer-Kurlbaum linear bolometer *T*. On rotating the mirror *C* about a vertical axis, the radiations of different wavelengths fall on the bolometer one after the other. The deflection of the galvanometer gives the corresponding spectral radiancy $E\lambda$, which is defined such that the quantity $E\lambda\ d\lambda$ is the energy, for wavelengths lying between A and $\lambda + d\lambda$, emitted per second per unit surface area of the black body.

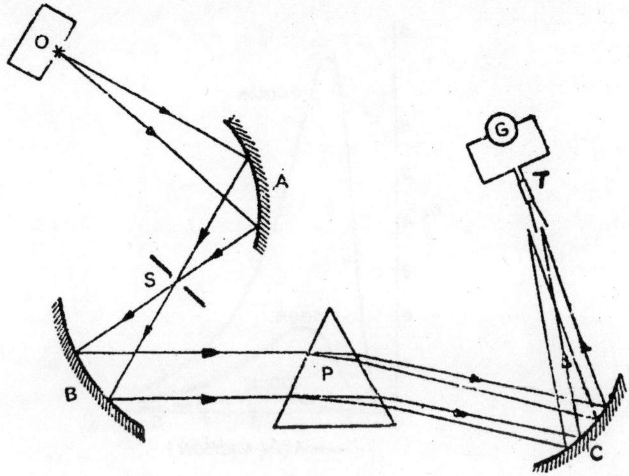

Fig. shows observed spectral radiancy $E\lambda$ plotted against the wavelength A of radiation at three different black-body temperatures. These curves show three important features :

· (*i*) The total energy emitted per second per unit area (*i.e.* radiancy E or area under the curve) increases rapidly with increasing temperature T. THE increase is found in accordance with the Stefan's law :

$$E = \int_0^\infty E_\lambda \, d\lambda = \sigma T^4.$$

(*ii*) At a particular temperature, the spectral radiancy E_λ is a maximum at a particular wavelength λ_m (say). Most of the energy is emitted at wavelengths not very different from λ_m

(*iii*) The wavelength λ_m for maximum spectral radiancy decreases in direct proportion to the increase in temperature. This is called 'Wien's displacement law' According to this law,

$$\lambda_m \times T = \text{constant.}$$

This means that as the temperature is raised, the cavity emits more and more the radiation of the shorter wavelengths.

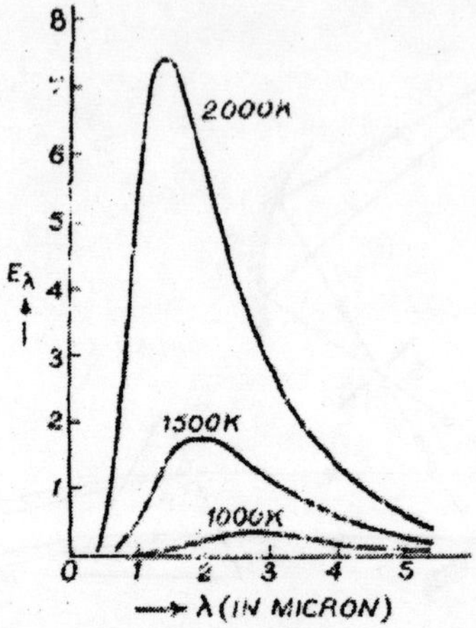

Success and Limitations of Classical Theory

Wien's and Rayleigh-Joans Formulae: First of all, Wien in 1893, worked on the distribution of energy among the wavelengths. He showed from pure thermodynamics reasoning that the energy-density $u\lambda$ in the wavelength-interval λ to $\lambda + d\lambda$ emitted by a black body at temperature T is of the form

$$d\lambda\ d\lambda = \frac{A}{\lambda^5}\ \mathrm{f}\ (\lambda T)\ d\lambda,$$

where A is a constant and $f\ (\lambda T)$ is an undetermined function. This is in agreement with the experimental result that the product λT at the peak of the $F\lambda$-λ curve is same at all temperatures.

In order to find the form of $f\,(\lambda T)$, Wien assumed that the black-body radiation inside a cavity may be supposed to be emitted by resonators of molecular dimensions having. Maxwellian velocity distribution, and the frequency of emitted radiation is proportional to the kinetic energy of the corresponding resonator. On this basis, Wein established the following distribution formula :

$$u\lambda\, d\lambda = \frac{A}{\lambda^5}\, e^{-B/\lambda T}\, d\lambda,$$

where A and B are constants.

Wein's formula was found to agree with experiment at short wave-lengths but did not fit well at long wavelengths (Fig.). According to the formula $u\lambda = 0$ for $\lambda = 0$ and also for $\lambda = \infty$ as it should be, but it keeps $u\lambda$ finite for $T = \infty$ which is unlikely.

Rayleigh and Jeans considered the black-body radiator (cavity) full of electromagnetic waves of all wavelengths between 0 and ∞ which, due to reflection at the walls, form standing waves. They calculated the number of possible waves having wavelengths between A and $\lambda + d\lambda$, and using law of equipartition of energy, established the following distribution formula :

$$u\lambda\, d\lambda = \frac{8\, \pi\, k\, T}{\lambda^4}\, d\lambda,$$

where k is the Boltzmann's constant,

Rayleigh-Jeans formula was found to agree with experiment at long wavelengths only (Fig.). It is, however, open to a serious objection. As we move towards shorter wavelengths (*i.e.* towards ultraviolet), the predicted energy would increase *without limit,* thus diverging enormously from experiment. This completely erroneous prediction is known as the 'ultraviolet catastrophe;'.

Furthermore, at any temperature T the total energy of radiation, $\int_0^\infty \frac{8\, \pi\, k\, T}{\lambda^4}\, d\lambda$, is predicted to be infinite which is against the Stefan's law.

The shortcomings of classical theory were overcome by Planck's quantum hypothesis is given below.

Explanation of Planck's hypothesis of quantum theory of radiation. Let's deduce an expression for the average energy of a Planck's oscillator and obtain Planck's radiation formula, the Rayleigh-Jean's law and Wien's law are special cases of Planck's law is proved here.

Planck's Quantum Hypothesis

Planck, in 1900, intro-entirely new ideas to explain the distribution of energy among the various wavelengths of the cavity radiation. He assumed that the *atoms of the walls of the cavity radiator behave as oscillators, each with a characteristic frequency of oscillation.* These oscillators emit electromagnetic radiant energy into the cavity and also absorb the same from it, and maintain an equilibrium state. Planck made two rather revolutionary assumptions regarding these atomic oscillators:

(*i*) *An oscillator can have only discrete energies given by*

$$\in \; = \; nh\nu,$$

where ν is the frequency of the oscillator, *h* is a constant known as 'Planck's constant', *n* is an integer known as 'quantum number'. This means that the oscillator can have only the energies $h\nu$, $2h\nu$, $3h\nu$,...and not any energy in between. In other words, the energy of the oscillator is quantised.

(*ii*) *The oscillators do not emit or absorb energy continuously but only in 'jumps'*. That is, an oscillator emits or absorbs packets of energy, each packet carrying an amount of energy $h\nu$.

$$\Delta \in \; = \; (\Delta n)\, h\nu \; ; \; \Delta n \; = 1.\ 2,...$$

Average Energy of Planck's Oscillator: Let us now calculate the average energy of Planck's oscillator of frequency ν. The relative probability that an oscillator has the energy $h\nu$ at temperature T is given by the Boltzmann, factor $e^{-hv/kT}$. Now, let $N_0, N_1 N_2 ... N_r$..be the number of oscillators having energies 0, hv, $2hv ... rhv$...respectively. Then we have $N_r = N_0 e^{-hv/kT}$. The total number of oscillators is

$$N = N_0 + N_1 + N_2 + \dots \dots$$

$$= N_0 (1 + e^{-hv/kT} + e^{-2hv/kT} + \dots \dots)$$

$$= \frac{N_0}{1 - e^{-hv/kT}}. \qquad ...(i)$$

The *total* energy of the oscillators is given by

$$\in = (N_0 \times 0) + (N_1 \times hv) + (N_2 \times 2hv) + ...$$

$$= (N_0 \times 0) + (N_0 e^{-hv/kT} \times hv) + (N_0 e^{-2hv/kT} \times 2hv) + \dots \dots$$

$$= N_0 e^{-hv/kT} hv(1 + 2e^{-hv/kT} + 3e^{-2hv/kT} + \dots \dots)$$

$$= N_0 e^{-hv/kT} \frac{hv}{\left(1 - e^{-hv/kT}\right)^{2*}} \qquad ... (ii)$$

Dividing eq. (ii) by eq. (i), we obtain the *average* energy of an oscillator as given by

$$\bar{\in} = \frac{\in}{N} = \frac{e^{-hv/kt}}{1 - e^{-hv/kT}} = \frac{hv}{e^{hv/kT-1}}. \qquad ...(iii)$$

This is the expression for the average energy of a Planck's oscillator.

Planck's Radiation Formula: The energy-density of radiation u_v in the frequency range v to v + dv is related to the average energy of an oscillator emitting v-radiation by

$$u_y d_v = \frac{8 \pi v^2}{c^3} dv \times \bar{\in}.$$

Substituting the value of $\bar{\in}$ from eq. (iii), we have

$$u_v dv = \frac{8 \pi v^2}{c^3} \frac{hv}{e^{hv/kT} - 1} dv$$

or $$u_v\, dv = \frac{8\pi h}{c^3} \frac{v^3\, dv}{e^{hv/kT} - 1}.$$

This is Planck's radiation formula in terms of frequency v to express it in terms of wavelength we observe that, since

$v = \frac{c}{\lambda}$,

$$dv = -\frac{c}{\lambda^2}\, d\lambda$$

and since an increase in frequency corresponds to a decrease in wavelength,

$$u_\lambda\; d\lambda = -\, u_v\, dv$$

Therefore $$u_\lambda\; d\lambda = \frac{8\pi h}{c^3} \frac{\left(\frac{c}{\lambda}\right)^3 \left(\frac{c}{\lambda^2}\, d\lambda\right)}{e^{hc/\lambda kT} - 1}$$

or $$u_\lambda\; d\lambda = \frac{8\pi hc}{\lambda^5} \frac{1}{e^{hc/\lambda kT} - 1} \qquad \text{... (iv)}$$

This is Planck's formula in terms of wavelength.

Explanation of Energy Distribution by Planck's Formula: Wien's Law and Rayleigh-Jeans Law are Special Cases : The Planck's formula is found to be in complete agreement with experiment for the entire wavelength range at all temperatures. This can be seen in the following way :

(1) When λ is very small, then $e^{hc/\lambda kT} >> 1$, so that Planck's formula given by eq. (IV) can be written as

$$u_\lambda\; d\lambda = \frac{8\pi hc}{c^5}\, e^{-\,hc/\lambda kT}\, d\lambda.$$

Putting $8\,\pi hc = A$ and $\frac{hc}{k} = B$, we have

$$u_\lambda\; d\lambda = \frac{A}{\lambda^5}\, \mathrm{e}^{-B/\lambda T}\, d\lambda.$$

This is Wien's law which agrees with experiment at short wavelengths.

(2) When λ is very large, then $e^{hc/kT} \approx 1 + \frac{hc}{\lambda kT}$ so that (iv) can be written as

$$u_\lambda \; d\lambda = \frac{8 \pi hc}{\lambda^5 \left(1 + \frac{hc}{\lambda kT} - 1\right)} d\lambda$$

$$= \frac{8 \pi k T}{\lambda^4} \; d\lambda$$

This is Rayleigh-Jeans law which agrees with experiment at long wavelengths.

Further, both Wien's displacement law and Stefan's law can be derived from the Planck's formula.

Wien's Displacement Law from Planck's Formula: The Planck's radiation formula is

$$u_\lambda \; d\lambda = \frac{8 \pi hc}{\lambda^5} \frac{d\lambda}{e^{hc/\lambda kT} - 1}$$

or $$u_\lambda = 8\pi hc \; (\lambda^{-5}) \; (e^{hc/\lambda kT} - 1)^{-1}.$$

To find the wavelength at which the spectral radiancy is maximum, we put

$$\frac{du_\lambda}{d\lambda} = 0$$

that is

$$8\pi hc \left[-5(\lambda^{-6}) \, (e^{hc/\lambda kT} - 1)^{-1} + \lambda^{-5} \, (-1) \, (e^{hc/\lambda kT} - 1)^{-2} \, e^{hc/\lambda kT} \frac{-hc}{\lambda^2 kT}\right] = 0$$

or $$\frac{5}{\lambda} = (e^{hc/\lambda kT} - 1)^{-1} \, e^{hc/\lambda kT} \frac{hc}{\lambda^2 \, kT}$$

or $$5 = \frac{hc}{\lambda kT} \frac{e^{hc/\lambda kT}}{e^{hc/\lambda kT} - 1}$$

Putting $\frac{hc}{\lambda kT} = x$, we get

$$5 = \frac{xe^x}{e^x - 1} = \frac{x}{1 - e^{-x}}$$

or $$\frac{x}{5} + e^{-x} = 1.$$

This equation has a single root given by $x = 4.965$, and therefore x must be a constant. That is

$$\frac{hc}{\lambda kT} = 4.965$$

Therefore, the wavelength λ_m at which the spectral radiancy per unit range of wavelength has its maximum value is given by

$$\lambda_m T = \frac{hc}{4.965\,k} = b \text{ (say)}$$

This is Wien's displacement law.

Stefan's Law from Planck's Formula: Let us write the Planck's radiation formula in terms of frequency ;

$$u_v\, dv = \frac{8\pi h}{c^3} \frac{v^3\, dv}{c^{hv/kT} - 1}$$

The spectral radiancy E_v is related to the energy-density u, by

$$E_v = \frac{c}{4} u_v$$

Thus $$E_v\, dv = \frac{2\pi h}{c^2} \frac{v^2\, dv}{e^{hv/kT} - 1}$$

The total radiant energy over all frequencies is

$$E = \int_0^\infty E_v\, dv = \frac{2\pi h}{c^2} \int_0^\infty \frac{V^3\, dv}{e^{hv/kT} - 1}$$

Let us put $\frac{hv}{kT} = x$, so that $v = \frac{Kt}{h}x$ and $dv = \frac{Kt}{h}dx$. Then

$$E = \frac{2\pi k^4 T^4}{h^3 c^2} \int_0^\infty \frac{x^3\, dx}{e^{x-1}}.$$

The value of the integral $\int_0^\infty \frac{x^3\, dx}{e^{x-1}}$ is $\frac{\pi^4}{15}$. Thus

$$E = \frac{2\,\pi^5 k^4}{15h^3c^2} T^2$$

Let us put $\frac{2\,\pi^5 k^4}{15h^3c^2} = \sigma$ (a universal Constant). Then

$$E = \sigma T^4.$$

This is Stefan's law.

Photon

The different forms of energy such as radio waves, heat rays, ordinary light, *X*-rays, *y*-rays, etc., which are emitted by atoms under different situations are all electro-magnetic radiations varying in frequency (or wavelength). We call the electro-magnetic *waves* because under suitable circumstances they exhibit refraction, interference and diffraction.

The wave-like character of radiation, however, failed to explain the observed energy distribution in the continuous spectrum of radiation emitted by hot bodies. To meet this serious problem, Planck, in 1901, presented his quantum theory of radiation According to this theory, radiation is emitted (or absorbed) *discontinuously* in *indivisible* packets of energy. These packets were named as 'photons' or 'Quantas. Each photon of radiation of a given frequency v has the same energy which is *hv*, where *h* is now known as Planck's constant. Planck did not suggest anything new regarding the propagation of radiation in space.

Einstein extended Planck's quantum hypothesis by assuming that *radiation not only is emitted (or absorbed) as indivisible photons, but also continues to propagate through space as photons.* On this hypothesis he in 1905 explained photoelectric effect and in 1907 tackled the problem of specific heat of solids. In 1913, Bohr used the quantum theory to explain the hydrogen spectrum and in 1922, Compton applied it to the scattering of X-rays.

Properties of Photons

(i) Photons are *indivisible* packets of electromagnetic energy.

(ii) They retain their identity until completely absorbed by some atom (as happens in photoelectric effect and pair production).

(iii) The size (energy content) of a photon (hv) is proportional to the frequency of radiation so that photons of different radiations are of different sizes. For example, blue photons are larger than red photons; X-ray photons are considerably larger than visible light photons.

(iv) The intensity of radiation (I) is equal to the number of photons (N) crossing unit area per second multiplied by the size (hv) of the photon *i.e.,*

$$I = N\,hv.$$

Thus for a given frequency the intensity depends simply upon the number of photons.

(v) *All* photons travel with the speed of light in vacuum (,) and have a zero rest mass. Hence there is no frame in which a photon is at rest.

(vi) From the special theory of relativity, the total energy E of a particle is related to its rest mass m_0 and momentum p by the relation

$$E = \sqrt{(m_0^2c^4 + p^2c^2)}.$$

For a Photon, $m_0 = 0$; therefore

$$E = pc.$$

But $E = h\nu$. Thus the photon has a momentum given by

$$p = \frac{E}{c} = \frac{h\nu}{c}$$

The Compton effect is a direct evidence of the existence of momentum for a photon.

(vii) The existence of momentum for a photon implies that it must have an effective mass m also. This mass can be computed by mass-energy relation ($E = mc^2$).

$$m = \frac{E}{c^2} = \frac{h\nu}{c^2}.$$

Photons and Corpuscles: The particle-like photon picture of radiation appears to be a return to the Newton's corpuscular theory of light. It, however, is not quite so. Corpuscles were material particles (with finite rest mass) whereas photons are energy packets with zero rest mass. Moreover, the energy of photons is related to the frequency of radiation. The photons, unlike corpuscles, include the wave picture of radiation also.

Complementarity of Waves and Corpuscles : In certain events radiation shows a wave-like character while in others it shows a particle-like character. The same light beam which is diffracted by a grating (showing wave-like character) can cause the emission of photo electrons from a suitable surface (showing particle-like character). These processes, however, occur independently. Thus the wave and particle aspects of radiation complement each other. Hence the radiation is supposed to have a dual character behaving as a wave in one situation and as a particle in the other situation,

PROBLEMS

1. Find the wavelength at which the spectral radiancy of a cavity radiator at 6000 K is maximum.

Solution : By Wien's displacement law, the wavelength λ_m for the maximum radiation is given by

$$\lambda_m T = \frac{hv}{4.965\, k}.$$

Putting $h = 6.62 \times 10^{-34}$ joule-sec, $c = 3.0 \times 10^8$ meters / sec and $k = 1.38 \times 10^{-23}$ joule / K, we get

$$\lambda_m T = \frac{\left(6.62 \times 10^{34}\right) \times \left(3.0 \times 10^{8}\right)}{4.965 \times \left(1.38 \times 10^{-23}\right)}$$

$$= 2.898 \times 10^{-3} \text{ Meter -K.}$$

For $T = 6000$ K, we get

$$\lambda_m = \frac{2.898 \times 10^{-3}}{6000}$$

$$= 0.483 \times 10^{-6} \text{ meter-K.}$$

$$= 4830 \text{ Å.} \qquad [1\text{A} = 10^{-10} \text{ meter}]$$

2. From Planck's law after deducing the value of v corresponding to the peak of the E_v— v curve for black body radiation at 1000 K. ($h = 6.62 \times 10^{-34}$ joule-sec, $k = 138 \times 10^{-23}$ joule/K) The result is as follows.

Solution : The wavelength λ_m at which the radiant energy is maximum, is given by

$$\lambda_m T = \frac{hc}{4.965\, k}$$

If v be the frequency corresponding to λ_m , then

$$v = \frac{c}{\lambda_m} = \frac{4.965\, kT}{h}$$

Here T = 1000 K.

$$\therefore \qquad v = \frac{4.965 \times \left(1.38 \times 10^{23} \text{ joule / K}\right)\left(1000 \text{ K}\right)}{6.62 \times 10^{-84} \text{ joule} - \text{sec}}$$

$$= 1.035 \times 10^{14} \text{ sec}^{-1}$$

3. The maximum in the energy distribution spectrum of the sun is at 4735 Å and its temperature is 6050 E. The temperature of a star whose energy distribution shows a maximum at 9506 Å will be as follows

Solution : By Wein' s displacement law,

$$\lambda_m T = \text{constant.}$$

Applying it for the sun and the star, we have

$$4735 \text{ Å X } 6050 \text{ K} = 9506 \text{ Å} \times T$$

or
$$T = \frac{4735 \text{ Å} \times 60.50 \text{ K}}{9506 \text{ Å}}$$

$$= 3013 \text{ K.}$$

4. A black body at 400 K radiates energy at the rate of 1.45 × 10³ wstt/m². We can try to obtain a value for Stefan's constant.

Solution : By Stefan's law, we have

$$E = \sigma T^4$$

Substituting the given values of E and T, we have

$$\sigma = \frac{E}{T^4}$$

$$= \frac{1.45 \times 10^2 \text{ watt / m}^2}{\left(400 \text{ K}\right)^4}$$

$$= 5.66 \times 10^{-8} \text{ watt/(m}^2 — \text{K}^4\text{).}$$

5. The surface temperature of the star which is radiating 1.6×10^6 times more energy per unit area per second compared to sun may be evaluated here. Surface temperature of the sun is 6000 K.

Solution : Let T be the surface temperature of the star. Then by Stefan's law, we have

$$1.6 \times 10^5 = \frac{T^4}{(6000\text{ K})^4}$$

or

$$T^4 = 1.6 \times 10^5 \times (6000\text{ K})^4$$

$$= 16 \times 36 \times 36 \times 10^{16}$$

$$\therefore \qquad T = 120000\text{ K}.$$

6. Let us calculate the energy of a photon of ultraviolet light of wavelength 1800 A and of infrared light of wavelength 36 μ, (h = 6.63 ×10^{-27} erg sec, c = 30 × 10^{10} cm/sec and 1 eV = 1.6 × 10^{-12} erg).

Solution : The energy of a photon of frequency v is

$$E = hv = h\frac{c}{\lambda},$$

where c is the velocity of light and λ the wavelength. Here λ = 1800 Å = 1800 × 10^{-8} cm.

$$\therefore \qquad E = \frac{\left(6.63 \times 10^{-27}\text{ erg} - \text{sec}\right) \times \left(3.0 \times 10^{10}\text{ cin sec}^{-1}\right)}{1800 \times 10^{-8}\text{ cm}}$$

$$= 1.1 \times 10^{-11}\text{ erg}$$

$$= \frac{1.1 \times 10^{-11}}{1.6 \times 10^{-12}}$$

$$= 6.9\text{ eV}.$$

For $\lambda = 3.6\ \mu = 3.6 \times 10^{-4}$ cm, we can see that

$$E = 0.345\text{ eV}.$$

7. The wavelength of a 100-MeV photon. (h = 6.63 × 10^{-34} joule-sec, c = 3 × 10^8 meters-sec-^{1}, and 1 eV = 1.6 ×10^{19} joule.) may be calculated

Solution : The energy of the photon is given by

$$E = hv = \frac{hc}{\lambda}$$

$$\therefore \qquad \lambda = \frac{hc}{E}$$

Here E = 100 MeV = 100 × 10^c eV = 10^8 × (1.6 × 10^{-10}) joule.

$$\therefore \qquad \lambda = \frac{\left(6.63\times10^{-34}\text{ joule-sec}\right)\times\left(3\times10^{8}\text{ meters sec}^{-1}\right)}{10^{8}\times\left(1.6\times10^{19}\right)\text{joule}}$$

$$= 124 \times 10^{-14} \text{ meter} = 1.24 \times 10^{-4} \text{ Å}$$

8. We can calculate in A the wavelength of a photon whose quantum energy is equal to the rest energy of an electron (m_0 = 9.1 × 10^{-31} kg, h = 6.62 × 10^{-34} joule-sec, c = 3 × 10^8 meter/ sec).

Solution : The rest energy of electron is w_0c^2. The wave-length of a photon whose energy is E = w_0c^2 is given by

$$\lambda = \frac{hc}{E} = \frac{h}{m_0c}$$

Putting the given values :

$$\lambda = \frac{6.62\times10^{-34}}{\left(9.1\times10^{-31}\right)\times\left(3\times10^{8}\right)}$$

$$= 0.024 \times 10^{-10} \text{ meter} = 0.024 \text{ Å}.$$

9. A faint star is just visible if 2000 photons/sec enter the eye. What energy per second (in watts) this represents in terms of sodium light is given below (λ = 5890 Å). (h = 6.63 × 10^{-34} joule-sec and c = 3 × 10^8 m/sec).

Solution : The energy of a photon is

$$E = h\nu = \frac{hc}{\lambda}.$$

Here $\lambda = 5890\ \text{Å} = 5890 \times 10^{-10}$ meter.

$$\therefore \quad E = \frac{(6.63 \times 10^{-34}) \times (3 \times 10^{8})}{5890 \times 10^{-10}}$$

$$= 3.38 \times 10^{-19} \text{ joule.}$$

The number of photons entering the eye is 2000 per sec. Hence the total energy received

$$= 2000 \times (3.38 \times 10^{-19})$$

$$= 676 \times 10^{-10} \text{ joule/sec}$$

$$= 6.76 \times 10^{-16} \text{ watt.}$$

10. A 1000-watt radio transmitter operates at a frequency of 880 kc/sec. We can see as to how many photons per second it emits ($h = 6.63 \times 10^{-34}$ joule-sec).

Solution : The electromagnetic energy radiated by the 1000-watt transmitter is 1000 joule/sec. The energy of a photon of radiation of frequency ν (880×10^3/sec) is

$$E = h\nu = (6.63 \times 10^{-34} \text{ joule-sec}) \times (880 \times 10^3 \text{ sec}^{-1})$$

$$= 5.83 \times 10^{-28} \text{ joule.}$$

Hence the number of photons emitted per sec

$$= \frac{\text{total energy emitted per sec}}{\text{energy of one photon}}$$

$$= \frac{1000 \text{ joules per sec}}{5.83 \times 10^{-28}} = 1.71 \times 10^{30} \text{ per sec.}$$

11. If a 60-watt light source is emitting light of wavelength 5400 A. The rate of emission of photons from the source would be as follows

Solution : The energy of a photon of light of wavelength $\lambda = 5400$ Å is

$$E = hv = \frac{hc}{\lambda}$$

$$= \frac{\left(6.63 \times 10^{34} \text{ joule-sec}\right) \times \left(3.0 \times 10^{8}\ m - sec^{-1}\right)}{\left(5400 \times 10^{10} \text{ m}\right)}$$

$$= 3.68 \times 10^{-19} \text{ joule.}$$

The rate of energy emission from a 60-watt lamp is 60 joules/sec. Hence the number of photons, each of energy 3.68×10^{-19} joule, emitted per second is

$$\frac{60}{3.68 \times 10^{-19}} = 1.63 \times 10^{20}.$$

12. Solar radiation falls on the earth at a rate of 1340 watt/ m² on a surface normal to the incoming rays. Assuming that sunlight consists exclusively of 5500 Å photons, the volume of photons per meter² per sec are reaching the earth is as follows

Solution : The energy of solar radiation is 1340 watts/m² or 1340 joules/m²-sec. The energy of a 5500-Å photon is

$$E = \frac{hc}{\lambda}$$

$$= \frac{\left(6.63 \times 10^{34} \text{ joule-sec}\right) \times \left(3.0 \times 10^{8}\ m - sec^{-1}\right)}{\left(5500 \times 10^{-10} \text{ m}\right)}$$

$$= 3.616 \times 10^{-19} \text{ joule.}$$

The number of these photons reaching the earth is

$$\frac{1340}{3.616 \times 10^{-19}} = 3.71 \times 10^{21} \text{ photons/m}^2\text{- sec.}$$

13. Radiation of wavelength 5000 A and intensity 2×10^{-2} watt/cm² falls on a photosensitive surface. Assuming

that every photon is absorbed and results in the ejection of a photoelectron, the way of finding the number of photo electrons are produced per cm^2 per sec is as follows. (h = 6.63 × 10^{-27} erg-sec, c = 3 ×10^{10} cm/sec).

Solution : The intensity of radiation is 2×10^{-2} watt/cm^2 or $2 \times 10^{-2} \times 10^7$ ergs*/cm^2-sec. The energy of a 5000-Å photon is

$$E = \frac{hc}{\lambda}$$

$$= \frac{(6.63 \times 10^{-27} \text{ erg-sec})(3 \times 10^{10} \text{ cm} - \text{sec}^{-1})}{(5500 \times 10^{-8} \text{ cm})}$$

$$= 3.978 \times 10^{-12} \text{ erg.}$$

∴ no of photons

$$= \frac{\text{intensity of radiation}}{\text{photon energy}}$$

$$= \frac{2 \times 10^{-2} \times 10^7}{3.978 \times 10^{-12}}$$

$$= 5.03 \times 10^{16} \text{ / cm}^2 \text{ - sec.}$$

Since every absorbed photon is ejecting a photoelectron, the number of photo electrons produced per cm^2 per second is 5.03×10^{16}.

14. The frequency of an x-ray photon whose momentum is 1.1 × l0^{-18} gm-cm/sec is explained below (h = 6.63 × 10^{-27} erg-sec and c = 3 × l0^{10} cm-sec^{-1}.)

Solution : The energy of a photon of radiation of frequency v is hv, and its momentum p is hv/c. Here

$$p = \frac{hv}{c}$$

$$= 1.1 \times 10^{-18} \text{ gm-cm/sec.}$$

$$\therefore \quad v = \frac{c}{h} \times p$$

$$= \frac{3 \times 10^{10}}{6.63 \times 10^{-27}} \times (1.1 \times 10^{-28})$$

$$= 5 \times 10^{18} \text{ cycles/sec.}$$

15. An x-ray photon has a wavelength 0.2 Å. Its energy (in eV) and momentum may be computed.

Solution : The energy E of a photon of frequency v is given by

$$E = hv = \frac{hc}{\lambda},$$

where λ is the wavelength.

Here A = 0.2 Å = 0.2 × 10⁻⁸ cm.

$$\therefore \quad E = \frac{(6.63 \times 10^{-27} \text{ erg} - \text{sec}) \times (3 \times 10^{10} \text{ cm} - \text{sec}^{-1})}{(0.2 \times 10^{8} \text{ cm})}$$

$$= 9.945 \times 10^{-8} \text{ erg}$$

$$= \frac{9.945 \times 10^{-8}}{1.6 \times 10^{-12}} \qquad [\therefore 1 \text{ eV} = 1.6 \times 10^{-12} \text{ erg}]$$

$$= 6.2 \times 10^{4} \text{ eV.}$$

The momentum p of a photon of frequency v is given by

$$p = \frac{hv}{c} = \frac{E}{c}$$

$$= \frac{9.94 \times 10^{-8} \text{ erg}}{3 \times 10^{10} \text{ cm - sec}^{-1}}$$

$$= 33.15 \times 10^{-18} \text{ gm-cm/sec}$$

17. The effective mass of a photon of wavelength 10 Å ($h = 6.62 \times 10^{-34}$ joule-sec and $c = 3 \times 10^8$ m/sec) may be computed.

Solution : A photon is a particle of zero rest mass moving at the speed of light and having a finite energy given by

$$E .= mc^2.$$

Its effective' mass *in* is therefore

$$m = E/c^2.$$

For a photon of wavelength. ;

We have $E = hv = hc/\lambda$.

$$\therefore \quad m = \frac{h}{c\lambda}$$

$$= \frac{6.62 \times 10^{-34} \text{ joule - sec}}{\left(3 \times 10^8 \text{ m / sec}\right) \times \left(1.0 \times 10^{-10} \text{ m}\right)}$$

$$= 2.21 \times 10^{-22} \text{ kg.}$$

8

The Two Effects

Photoelectric Effect

When a beam of light of frequency in the blue or ultraviolet region falls on a metal plate, blow-moving electrons are emitted from the metal surface. This phenomenon is known as 'photoelectric effect' and the electrons emitted are known as 'photoelectrons'. The photoelectrons can be obtained from non-metals and gases also, provided that light of the appropriate frequency is used.

Experimental Observation: Lenard in 1900 studied the photoelectric effect experimentally. His apparatus (Fig.) consists of an evacuated tube having two plates, *C* and *A*. A varying potential difference can be applied between C and *A* by means of a potential divider and a commutator *K*. The p.d. and the current in the circuit can be read by a voltmeter and an ammeter respectively.

Dependence Upon Intensity: Light of an appropriate *fixed* frequency is allowed to fall on the plate *C*. Photoelectrons are emitted from the surface of C. When the potential V of the plate A is made positive relative to C, a steady saturation current flows in the circuit, whatever the (positive) value of V. This means that all the photoelectrons emitted from C are reaching A. If the intensity of the incident light is increased, the saturation current is found to increase in the same ratio. This shows that the number of photoelectrons emitted per second varies directly as the intensity of the incident light.

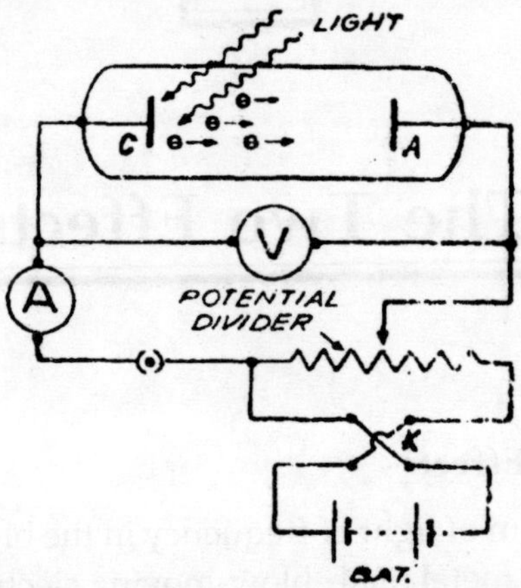

When the potential V of the plate A is made negative with respect to C, the electrons are retarted. If the potential is small, only the slower electrons are pulled back and the current falls. As the negative potential is increased, more and more faster electrons are pulled back. Ultimately, at a certain potential V_0, even the fastest electrons fail to reach A. The photoelectric current is then zero. An increase, however large, in the intensity of the incident light does not produce any current at the potential V_0 (Fig.). The (negative) potential V_0 *at* which the current is just reduced to zero is called the 'stopping potential. Since the energy of the fastest electrons is just annuled when they fall through the stopping potential, the stopping potential is a measure of the maximum kinetic energy of the photoelectrons. Hence we conclude that *the maximum energy of the photoelectrons is independent of the intensity of the incident light.*

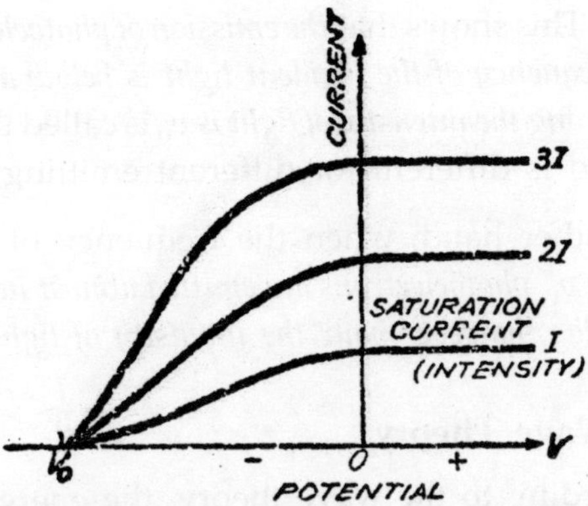

Dependence Upon Frequency : If, after reaching at the stopping potential, the incident light is replaced by another light of higher frequency, once more a current begins to flow in the circuit. This current again stops when the stopping potential is increased. Thus higher the frequency of the incident light, the higher is the stopping potential (Fig.) *i.e.* higher the maximum energy of the photo electrons

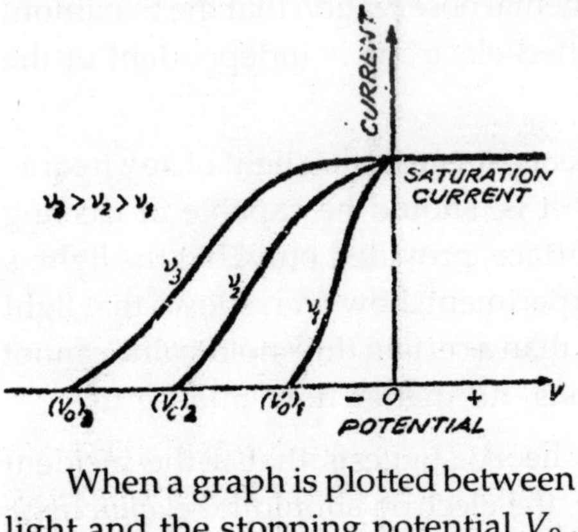

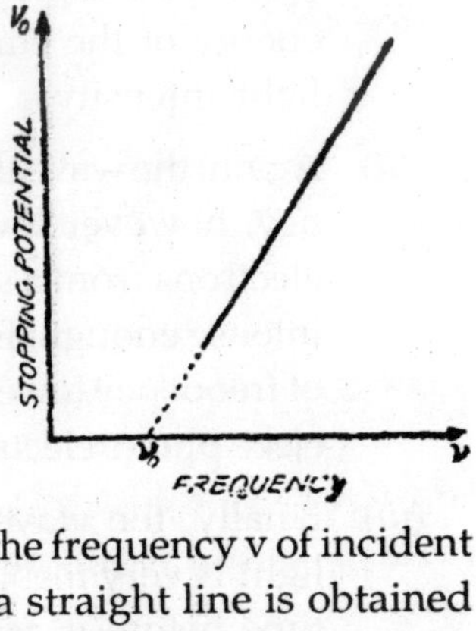

When a graph is plotted between the frequency v of incident light and the stopping potential V_0, a straight line is obtained (Fig.). From this we conclude that *the maximum energy of the photoelectrons increase; linearly as the frequency of the incident tight increases* .

If the straight-line graph (between V_0 and v) is produced back, it cuts the v-axis at a point corresponding to a particular frequency v_0. This shows that *the emission of photoelectrons cannot occur if the frequency of the incident light is below a certain value* v_0, *however strong the intensity of light is* v_0 is called the 'threshold frequency and is different for different emitting surfaces.

On the other hand, when the frequency of the incident light is above v_0, *photoelectrons are emitted almost instantaneously with no time-lag, however pour the intensity of light.*

Failure of Wave Theory

(*i*) According to the wave theory, the energy carried by a light beam is distributed uniformly and continuously over a wavefront *and is measured by the intensity of the beam.* Thus when light fails on a metal surface the energy of the wave should be transferred uniformly to the electrons in the surface before they are emitted out. Obviously, the energy taken up by the electrons must increase as the intensity of light is increased. This is against the experimental observation that the maximum energy of the emitted electrons is independent of the light intensity.

(*ii*) Again, the wave theory suggests that light of any frequency, however low it is, should be capable of ejecting electrons from a surface, provided only that the light is intense enough. Experiment, however, shows that light of frequency lower than a certain threshold value cannot eject photo electrons, no matter how intense it is.

(*iii*) Finally, the wave theory suggests that if the incident light is very feeble, the electron should take *appreciable* time before it acquires sufficient energy to come out from the surface. However, no detectable time-lag has ever been found between the falling of the light on the surface and the emission of the photo electrons. Thus wave theory fails in explaining the experimental observations regarding the photoelectric effect.

Explanation from Quantum Theory

According to the quantum theory, the energy carried by a light beam of frequency v is concentrated in indivisible packets called 'photons', each photon carrying an amount *hv*. The intensity of the beam depends merely upon the number of photons. Now when light falls on a metal surface, a photon (energy *hv)* is *completely* absorbed by a single electron in the surface. Part of this energy is used in ejecting the electron against the attraction of the rest of the metal and rest is given to the electron as kinetic energy. If the electron doss not lose energy by internal collisions, it emits out ,with a maximum kinetic energy. If the intensity of light is doubled, the number of photons falling per second on the surface will be doubled, the photon energy *hv* will remain the same) and hence the number of electrons emitted per second will and be doubled, their maximum energy remaining the same as before. Thus the number of photoelectrons emitted per second varies directly as the light intensity. But their maximum energy is independent of the intensity. This is completely is agreement with experiment

The maximum kinetic energy of the photoelectrons increases as the frequency of light, and hence the photon energy increases as observed by experiment.

The existence of a threshold frequency also speaks is favour of the photon (quantum) theory. The electron cannot leave the metal until the energy given to it by a single photon (*hv*) exceeds the energy required in liberating the electron against the attraction of the rest of the metal; no matter how many photons there are (that is no matter how intense the incident light is).

Finally, the absence of a time-lag also follows from the photon theory. As soon as the first photon strike the surface, in is immediately absorbed by *some* electron which escapes instantaneously, there being no time-lag.

The successful explanation of photoelectric effect in terms of quantum theory is one of the strongest evidences in favour of this theory.

Laws of Photoelectric Emission

There are following laws of photoelectric emission which have been derived from experimental observations :—

(1) The number of photoelectrons emitted per second varies directly as the intensity of the incident light.

(2) The maximum energy of the photoelectrons is independent of the intensity of the incident light.

(3) The maximum energy of the photoelectrons increases linearly as the frequency of the incident light increases.

(4) The emission of photo electrons cannot occur if the frequency of the incident light is below a certain value v_o. (called the threshold frequency), however strong the intensity of light, v_0 is different for different emitting surfaces.

(5) Photoelectrons are emitted almost instantaneously, however poor the intensity of the incident light is.

Einstein's Photon Theory and the Photoelectric Equation: Einstein, in 1905, explained the photoelectric effect by introducing the photon theory of light. He assumed that the light (radiation) travels through pace in *indivisible* packets of energy called photons' or 'quanta', the energy of a single photon being hv, where h is the Planck's constant' and v the frequency of the light.

When light falls on a metal surface, a photon (energy hv) is completely absorbed by a single electron in the surface. Part of this energy is used in electing the electron against the attraction of the rest of the metal and the rest is given to the electron as kinetic energy. Those electrons which are ejected from some depth below the surface expend some of their acquired energy in collisions with the atoms. The energy lost in this way is a variable quantity and so the photo electrons are emitted with a range of kinetic energies.

Let us consider those photoelectrons which are emitted from very near the surface and do not suffer and collision. They emit with maximum kinetic energy $(\frac{1}{2}mv^2)_{max}$. Let W be the energy required to eject an electron against the attraction of the rest of the metal. W is called the work function' of the given metal. Thus we can write

$$hv = W + \left(\frac{1}{2}mv^2\right)_{\text{max}} \qquad \dots (i)$$

hv being the photon energy absorbed by the electron. Now, an electron cannot emit from the surface if the energy of the photon absorbed by it is less than W. If v_0 be the threshold frequency, then the threshold energy of the photon will be hv_0. Thus

$$W = hv_o$$

Making this substitution in eq. (*i*), we get

$$hv = hv_0 + \left(\frac{1}{2}mv^2\right)_{max}$$

or

$$\left(\frac{1}{2}mv^2\right)_{\text{max}} = h(v - v_0). \qquad \dots (ii)$$

This is Einstein's photoelectric equation.

Explanation of the Photoelectric Laws

If the intensity of light of a given frequency v is increased, the number of photons striking the surface per second will increase, but the energy hv of each photon will remain the same. Hence the number of electrons emitted per second will correspondingly increase, but their maximum, energy $(\frac{1}{2}mv^2)_{\text{max}}$ will remain the same, as is clear from Einstein's equation. Thus the laws (1) and (2) are explained.

It is also seen from Einstein's equation that the maximum energy of the photoelectrons, $(\frac{1}{2}mv^2)_{max}$, will increase linearity with the increase in the frequency v of the incident light. This is the photoelectric law (3).

The eq. (*ii*) further shows that if $v<v_0$, the kinetic energy of the photo electrons would be negative which is impossible. This means that photo electrons cannot emit if the frequency v of the incident light is less than the threshold frequency v_0. This is law (4).

Finally, as soon as the first photon strikes the surface, it is immediately absorbed by some electron which escapes instantaneously, there being no time-lag. This is law (5).

In Einstein's theory the emission of one electron is associated with the absorption of one photon. This, however, does not mean that every photon incident on the metal causes emission of an electron even though the photon energy is above threshold. The photon can be involved in many alternative processes, and hence the ratio of the number of electrons emitted to the number of photons incident on the surface is very much less than unity.

Millikan's Verification: Einstein's equation was tested by R. A. Millikan who measured the maximum energies of the photoelectrons emitted by a number of alkali metals over a wide range of light frequencies. His apparatus is shown in Fig.

Three different alkali metal surfaces are mounted on a drum *D* placed inside an evacuated chamber. The drum can be rotated from outside so that any surface can be scraped clean by a sharp knife *K* and then brought before a window *W**. A beam of monochromatic light passing through the window falls on the surface C. The emitted photo electrons are collected by an electrode *A* by means of a potential difference applied between C and *A* through a potential-divider. The galvanometer *G* measures this photoelectric current.

When the electrode *A* is given a negative potential with respect to C, the electrons emitted from C are retarted. If the potential is small, only the slower electrons are pulled back and the current falls. As the negative potential is increased, more and more faster electrons are pulled back. Ultimately, at a certain potential V_0, even the fastest electrons fail to reach *A*. The photoelectric current is then zero. The (negative) potential V_0 at which the current is just reduced to zero is called the 'stopping potential. Since the energy of the fastest electrons is just annuled when they fall through the stopping potential, we have

$$(\tfrac{1}{2}mv^2)_{max} = eV_0\,, \qquad \dots (i)$$

where e is the electronic charge and V_0 is the stopping potential. Thus the stopping potential multiplied by electronic charge gives the maximum kinetic energy of the photoelectrons.

Millikan plotted a graph between the stopping potential and the frequency of light over a wide range of frequencies, and obtained a straight line (Fig.). Parallel lines were obtained for other metallic surfaces.

Now, according to Einstein's equation, we have

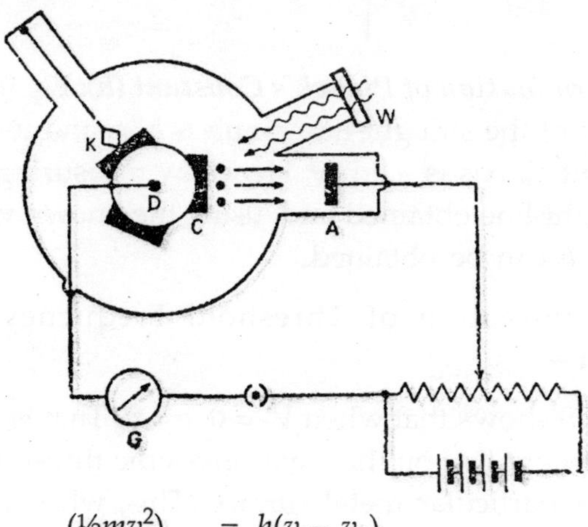

$$(\tfrac{1}{2}mv^2)_{max} = h(v - v_0).$$

But from (*i*), $(\frac{1}{2}mv^2)_{max} = eV_0$

$$\therefore \qquad eV_0 = h\,(v - v_0)$$

$$\text{or} \qquad V_0 = \left(\frac{h}{e}\right)v - \left(\frac{h}{e}\right)v_0 \ldots$$

Since h and e are constants and v_0 is constant for a given surface, eq. (*ii*) indicates that the graph between the stopping potential V_0 and the frequency of light v must be a straight line. This is actually the case, as found by Millikan. Hence Einstein's equation is of the correct form.

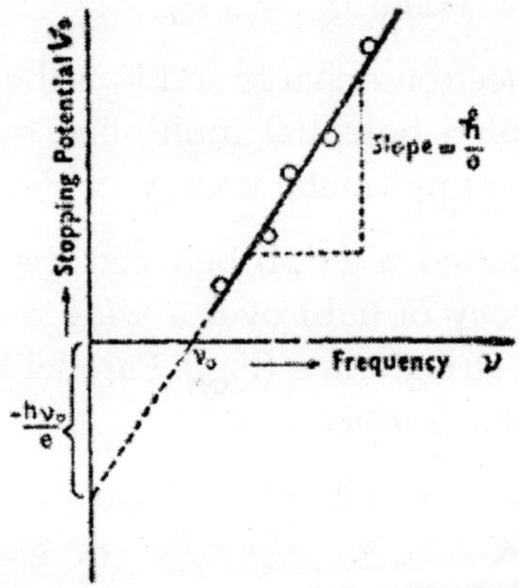

Determination of Planck's Constant (h): Eq. (*ii*) shows that the slope of the straight-line graph is h/e, and its intercept on the potential axis is $-hv_0/e$. Hence by measuring the slope of the straight-line obtained and, using the known value of e, the value of h can be obtained.

Determination of Threshold Frequency and Work Function —

Eq. (*ii*) shows that when $V_0 = 0$; $v = v_0$ That is, the intercept of the straight line on the v-axis gives the threshold frequency v_0 for the particular metal surface. This, when multiplied by h, gives the work function hv_0.

Let's see as to what is the effect on the energy and the number of emitted photoelectrons when (i) the intensity of incident light decreases, (ii) the target material is changed, (iii) the frequency of the incident light increases

The Einstein's photoelectric equation is

$$\left(\frac{1}{2}mv^2\right)_{max} = hv - W,$$

where v is the frequency of the incident light, W the work function of the metallic surface and $(½mv^2)_{max}$ the maximum kinetic energy (K_{max}) of the emitted electrons. If v_0 be the threshold frequency, then $W = hv_0$, and so

$$K_{max} = hv - hv_0.$$

A graph between K_{max} and v would be a straight line, cutting the v axis at v_0. The slope of this line is h.

(i) *Effect of Change in Light Intensity:* A or change in light intensity means a change in the number of photons falling per second on the metal. This will cause a corresponding change in the number of photoelectrons emitted. Thus if the light intensity is decreased, the number of photo electrons will decrease but their energy will remain the same. Hence the K_{max}–v graph will remain unchanged.

(ii) *Effect of Change of Target Material :* The work function that must be done to take an electron through the emitting surface from just beneath it against the abaction of the rest of the metal. Obviously, a change in metal would change the work function W (and hence the threshold frequency v_0) which will change the energy of the photoelectrons. The new graph will be another straight line *parallel* to the previous line. Parallel, be-cause slope (h) in fixed

(iii) *Effect of Change in Light Frequency :* A change in the frequency v of the incident light will cause a corresponding linear change in the maximum kinetic energy of the photoelectrons. A decrease in frequency (or increase in wavelength), for example, would decrease the energy. The number of photo electrons emitted per second will remain unaffected.

Compton Effect

When a *monochromatic* beam of X-rays is scattered by a substance, the scattered X-rays contain radiation not only of the same wavelength as that of the primary rays but also the radiation of longer wavelength. The former is called the 'unmodified radiation' and the latter the 'modified radiation'. The classical electromagnetic theory explained the unmodified radiation but it totally failed to explain the presence of the modified radiation.

Compton, however, gave a satisfactory explanation for the modified radiation on the basis of quantum theory. Hence the phenomenon giving rise to modified radiation is called 'Compton effect.

Theory: According to the quantum theory, the primary X-ray beam is made up of photons of energy hv, where h is Planck's constant and v is the frequency of the primary X–rays. These photons travel with the speed of light a and possess momentum given by hv/c..

Compton assumed that during the scattering process, the incident photon collides with a free electron (assumed to be initially at rest in the laboratory coordinate system) in the scattering material (Fig.a). The photon transfers some of its energy to the electron which recoils with a velocity v in a direction making an angle ϕ (say) with the direction of the incident photon (Fig. b). The photon itself *with reduced energy* is scattered in a direction θ with the original direction. These scattered photons constitute the scattered modified radiation.

Let us apply the laws of conservation of energy and momentum to this collision.

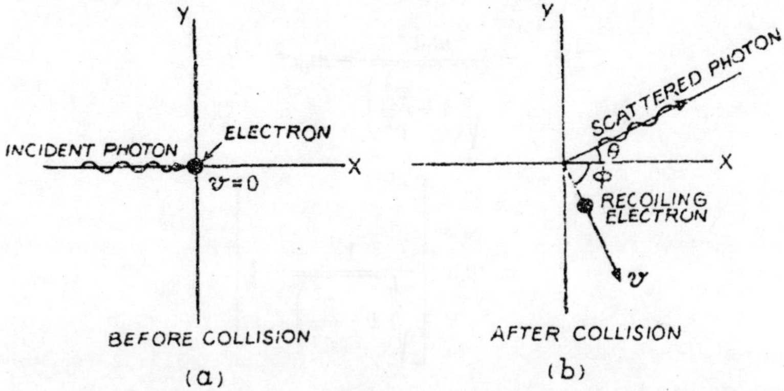

The energy and momentum of the incident photon are respectively hv and hv/c.

The energy and momentum of scattered photon are respectively hv' and hv'/c where v' is the frequency of the scattered radiation.

Since the recoiling electron may have a speed v which is comparable with the speed of light, we must use the relativistic expression for the energy and momentum of the electron. According to the theory of relativity, a mass m is equivalent to an amount of energy E by the relation

$$E = mc^2,$$

where c is the velocity of light. Further the mass of a particle varies with its velocity v according to the relation

$$m = \frac{m_0}{\sqrt{\left(1 - \frac{v^2}{c^2}\right)}}.$$

where m_0 is the mass of the particle when at rest. The total energy of the recoiling electron is the sum of its kinetic energy K plus its rest mass energy *i.e.*

$$mc^2 = K + m_0 c^2.$$

Hence the kinetic energy of the electron is

$$K = mc^2 - m_0c^2$$

$$= \frac{m_0c^2}{\sqrt{\left(1-\frac{v^2}{c^2}\right)}} - m_0c^2$$

$$= m_0c^2\left[\frac{1}{\sqrt{\left(1-\frac{v^2}{c^2}\right)}} - 1\right]$$

Similarly, the momentum of the recoiled electron is

$$mv = \frac{m_0v}{\sqrt{\left(1-\frac{v^2}{c^2}\right)}}$$

Now, applying the law of conservation of energy, we get

$$hv = hv' + m_0c^2\left[\frac{1}{\sqrt{\left(1-\frac{v^2}{c^2}\right)}} - 1\right]. \qquad \dots (i)$$

Again, applying the law of conservation of momentum along the X and Y axes, we have respectively

$$\frac{hv}{c} = \frac{hv'}{c}\cos\theta + \frac{m_0v}{\sqrt{\left(1-\frac{v^2}{c^2}\right)}}\cos\phi \qquad \dots (ii)$$

and

$$0 = \frac{hv'}{c}\sin\theta + \frac{m_0v}{\sqrt{\left(1-\frac{v^2}{c^2}\right)}}\sin\phi \qquad \dots (iii)$$

Re-arranging and squaring (*ii*) and (*iii*), we get

$$\left(\frac{hv}{c} - \frac{hv'}{c}\cos\theta\right)^2 = \frac{m_0^2 v^2}{1 - \frac{v^2}{c^2}}\cos^2\phi$$

and $$\left(\frac{hv'}{c}\sin\theta\right)^2 = \frac{m_0^2 v^2}{1 - \frac{v^2}{c^2}}\sin^2\phi$$

Adding : $$\left(\frac{hv}{c}\right)^2 + \left(\frac{hv'}{c}\right)^2 - \frac{2h^2vv'}{c^2}\cos\theta = \frac{m_0^2 v^2}{1 - \frac{v^2}{c^2}} \quad \dots (iv)$$

Re-arranging (*i*) and dividing by *c*, we get

$$\frac{hv}{c} - \frac{hv'}{c} + m_0 c = \frac{m_0 c}{\sqrt{\left(1 - \frac{v^2}{c^2}\right)}}$$

Squaring :

$$\left(\frac{hv}{c}\right)^2 - \left(\frac{hv'}{c}\right)^2 + m_0^2c^2 - \frac{2h^2vv'}{c^2} + 2hm_0\left(v - v'\right) = \frac{m_0^2c^2}{1 - \frac{v^2}{c^2}} \quad \dots (v)$$

Subtracting (*iv*) and (*v*), we get

$$m_0^2c^2 - \frac{2h^2vv'}{c^2}\left(1 - \cos\theta\right) + 2hm_0\left(v - v'\right) = \frac{m_0^2}{1 - \frac{v^2}{c^2}}\left(c^2 - v^2\right) = m_0^2c^2$$

$$\text{or } 2hm_0\left(v - v'\right) = \frac{2h^2vv'}{c^2}\left(1 - \cos\theta\right)$$

$$\text{or } \quad m_0\left(v - v'\right) = \frac{hvv'}{c^2}\left(1 - \cos\theta\right)$$

Substituting $v = c/\lambda$ and $v' = c/\lambda'$ where λ and λ' are the wavelengths of the incident and scattered *X*-rays, we get

$$m_0 c\left(\frac{1}{\lambda} - \frac{1}{\lambda'}\right) = \frac{h}{\lambda\lambda'}\left(1 - \cos\theta\right)$$

or $$\lambda' - \lambda = \frac{h}{m_0 c}(1 - \cos\theta).$$

or $$\Delta\lambda = \frac{h}{m_0 c}(1 - \cos\theta).$$

This is the expression for the change in wavelength, $\Delta\lambda$ is called the 'Compton shift'.

Compton Wavelength : The quantity h/m_0c has dimensions of length, and is known as the 'Compton wavelength of the electron'. Its magnitude is

$$\frac{h}{m_0 c} = \frac{6.63 \times 10^{-34} \text{ joule-sec}}{(9.11 \times 10^{-31} \text{ kg}) \times (3.0 \times 10^{8} \text{ meter-sec}^{-1})}$$

$$= 0.0243 \times 10^{-10} \text{ m} = 0.0243 \text{ Å}.$$

This shows that *the change in wavelength (Compton shift) depends neither on the incident wavelength nor on the scattering material,* but depends only on the angle of scattering.

At $$\theta = 90° ;$$

$$\Delta\lambda = 0.0243 \text{ Å}.$$

This result was verified by Compton himself.

Presence of the Unmodified Radiation: The above theory holds for the 'free' or 'loosely-bound' electrons of the atom. If the photon strikes an electron *'tightly bound'* to the atom, then the hole atom, and not simply the electron, recoils. Since the mass of the atom much larger, the Compton shift $\frac{h}{m_0 c}$ $(1 - \cos\theta)$, m_0 now being the mass of the atom, is negligible. This accounts for the unmodified radiation. In the heavier elements in which most of the electrons are tightly bound, the modified radiation is too weak.

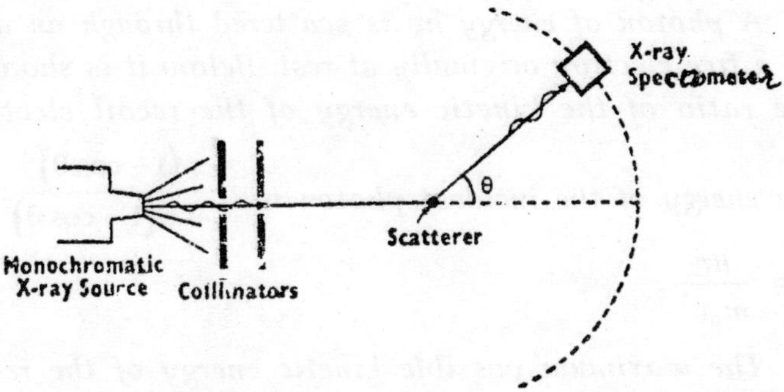

The Compton effect offers one of the strongest evidences in support of the Planck's quantum theory of radiation.

Experimental Verification: A beam of monochromatic X-rays of known wavelength λ is made to fall on a graphite scattered. The distribution of intensity with wavelength in the X-rays scattered at various angles θ is measured by means of a Bragg's X-ray spectrometer (Fig.). The results have been shown in. Fig. For $\theta = 90°$, $\Delta\lambda$ is 0.0242 Å, which is an agreement with Compton formula.

Measurement of Energy of the Recoiled Electrons : From eq. (*ii*), in which $v' < v$, it is clear that necessarily $\cos \phi > 0$ and hence $\phi < 90°$, so that the electron must move in the forward direction. These recoiled electrons, whose existence was predicted by Compton, were experimentally discovered by C. T. R. Wilson and Bothe. Bless measured the energies for these electrons recoiled in various directions with various velocities by using a magnetic spectrogram. The electrons are subjected to a magnetic; field in which they describe circles and are received on a photographic plate. Those electrons which leave different parts of the scanner with the same velocity describe circles of same radius and are focussed on the same line on the plate. By measuring the positions of the various lines on the plate (which represent groups of recoiled electrons), the radii of the corresponding circles are known from which the velocities are computed. Bless found that the results were in agreement with Compton's predictions.

A photon of energy $h\upsilon$ is scattered through an angle θ by a free electron originally at rest. Below it is shown that the ratio of the kinetic energy of the recoil electron to the energy of the incident photon is $\frac{x(1-\cos\theta)}{1+x(1-\cos\theta)}$ where $x = \frac{h\upsilon}{m_0c^2}$.

The maximum possible kinetic energy of the recoiling electron in terms of the incident photon energy may be computed.

The energy of incident photon is

$$E = h\upsilon.$$

Let υ' be the frequency of the scattered photon and K the kinetic energy of the recoil electron. By conservation of energy, we have

$$h\upsilon = h\upsilon' + K.$$

$$\therefore \quad \frac{K}{E} = \frac{h\upsilon - h\upsilon'}{h\upsilon} = \frac{(h\upsilon/\lambda) - (h\upsilon/\lambda')}{hc/\lambda}$$

$$= \frac{\lambda' - \lambda}{\lambda'} = \frac{\Delta\lambda}{\lambda + \Delta\lambda}.$$

The Compton shift $\Delta\lambda = \frac{m}{m_0c}(1-\cos\theta)$.

$$\therefore \quad \frac{K}{E} = \frac{(h/m_0c)(1-\cos\theta)}{\frac{c}{\upsilon} + \frac{h}{m_0c}(1-\cos\theta)}$$

$$= \frac{\frac{h\upsilon}{m_0c^2}(1-\cos\theta)}{1 + \frac{h}{m_0c^2}(1-\cos\theta)}$$

$$= \frac{x(1-\cos\theta)}{1+x(1-\cos\theta)} \cdot \quad \text{where } x = h\upsilon/m_0c^2.$$

The last equation may be written as

$$\frac{K}{E} = \frac{2x\sin^2\frac{\theta}{2}}{1+2x\sin^2\frac{\theta}{2}}$$

$$K = \frac{E}{\dfrac{1}{2x\sin^2\frac{\theta}{2}}+1}$$

K is maximum when the denominator on the R.H.S. is minimum or $\sin^2\frac{\theta}{2}$ is maximum *i.e.* $\theta = 180°$.

$$\therefore \qquad K_{max} = \frac{E}{\dfrac{1}{2x}+1} = \frac{2\,Ex}{1+2x}.$$

Now $E = h\upsilon$ and $x = h\upsilon/m_0c^2$.

$$\therefore \qquad K_{max} = \frac{2h^2\upsilon^2/m_0c^2}{1+2h\upsilon/m_0c^2}.$$

Comparison of Compton Effect and Photoelectric Effect : Let us consider a Compton collision between a photon and a free electron (initially at rest). The photon has an energy $h\upsilon$ and a momentum $h\upsilon/c$. Suppose the photon transfers its entire energy (and momentum) to the electron who acquires a velocity υ. Then, from consideration of the conservation of energy and momentum, we have

$$h\upsilon = \tfrac{1}{2}m\upsilon^2$$

and

$$\frac{h\upsilon}{c} = m\upsilon,$$

where m is the mass of the electron. Eliminating $h\nu$ between these two equations we find that the velocity of the electron comes out to be

$$v = 2c,$$

that is, twice the velocity of light. This is impossible according to the relativistic mechanics. Hence we conclude that a *'free' electron cannot absorb 'all' the energy of a photon and conserve both energy and momentum.*

In photoelectric effect, an electron absorbs the incident photon *completely* before it leaves the photo-sensitive surface as a photoelectron. Obviously, this electron *cannot be free i.e.* it must be bound to an atom otherwise it could not absorb the photon completely. Hence we conclude that the photoelectric effect cannot occur for completely free electrons; the electrons must be bound to an atom.

Thus there is an important difference between the photoelectric effect and the Compton effect. In the photoelectric effect the photon completely disappears and all of its energy is given to the photoelectron. In Compton effect there is still a photon after the collision but its energy is less than that of the incident photon, only a part of the energy of the incident photon has been given to the Compton electron. Thus in Compton effect, the electron can be free.

Compton Effect not Observed for Visible Light: The change in wavelength due to the Compton effect is given by

$$\Delta\lambda = \frac{h}{m_0 c}(1 - \cos\theta) = 0.0243\,(1 - \cos\theta)\ \text{Å},$$

where θ is the scattering angle. Obviously, the change $\Delta\lambda$ is independent of the wavelength of the radiation suffering Compton effect, and depends only on the scattering angle θ. The greatest value of $(1 - \cos\theta)$ is 2 when $\theta = 180°$, so that the maximum wavelength-change possible is 0.0486 or roughly

0.05 Å only. This means that Compton effect can be detected only for those radiations whose wavelength is not greater than a few λ. For example, for λ = 5 Å, there is a maximum wavelength-change of 1 %, while for λ = 1 Å there is a 5% change. For visible light (λ ≃ 5000 Å) the maximum wavelength change (0.05 Å) is only about 0.001% of the initial wavelength which is undetectable.

PROBLEMS

1. The photoelectric work function of a metal is 2.061 eV. Calculate the threshold wavelength and frequency for the metal. (h = 6.6 × 10^{-34} joule-sec, e = 3.0 × 10^8 m/sec, 1 eV = 1.6 × 10^{-19} joule).

Solution : A radiation photon cannot emit photo electron from a metal if its energy is less than the work function W of the metal. If v_0 be the threshold frequency of radiation, the threshold energy of the photon will be hv_0. Thus

$$W = hv_0.$$

$$\therefore \qquad v_0 = \frac{W}{h} = \frac{2.061 \times 1.6 \times 10^{-19} \text{ joule}}{6.6 \times 10^{-34} \text{ joule - sec}}$$

$$= 5.0 \times 10^{14} \text{ sec}^{-1}.$$

The threshold wavelength is

$$\lambda_0 = \frac{c}{v_0} = \frac{3.0 \times 10^8 \text{ m - sec}^{-1}}{5.0 \times 10^{14} \text{ sec}^{-1}}$$

$$= 6 \times 10^{-7} = 6000 \text{ Å}.$$

2. The work function of sodium is 2.3 eV. Let's see what is the maximum wavelength of light that will cause the photoelectrons to be ejected. It would be eager to look into whether sodium shows a photoelectric effect for orange light, λ = 6800 Å (h = 6.626 × 10^{-34} J-s, c = 3.0 × 10^8 m/s).

Its answer is: 5400 Å, No)

3. The work function in electron-volts for a metal, given that the photoelectric threshold wavelength is 6800 Å is calculated below

Solution : Threshold wavelength λ_0 = 6800 Å = 6800 × 10^{-10} meter. If v_0 be the threshold frequency, then the work function for the metal is

$$W = hv_0 = \frac{hc}{\lambda_0}$$

$$= \frac{(6.6 \times 10^{-34} \text{ joule - sec}) \times (3.0 \times 10^{8} \text{ m - sec}^{-1})}{2800 \times 10^{-10} \text{ m}}$$

$$= 2.9 \times 10^{-19} \text{ joule.}$$

Now $1 \text{ eV} = 1.6 \times 10^{-19}$ joule.

$$\therefore \quad W = \frac{2.9 \times 10^{-19}}{1.6 \times 10^{-19}} = 1.8\text{eV}.$$

4. A photon of wavelength 3310 Å falls on a photo cathode and ejects an electron of maximum energy 3 × 10^{-12} erg. The work function of the cathode material is calculated here. (h = 6.62 × 10^{-27} erg-sec, c = 3 ×10^{10} cm/sec and 1 eV = 1.6 × 10^{-12} erg).

Solution : When a photon of frequency v falls on a photo-cathode of work function W, the maximum kinetic energy K_{max} of the photoelectrons is given by the Einstein photoelectric equation:

$$K_{max} = hv - W$$

or

$$W = hv - K_{max} = \frac{hc}{\lambda} - K_{max}$$

Substituting the given values, we get

$$W = \frac{(6.62 \times 10^{-27} \text{ erg - sec}) \times (3 \times 10^{10} \text{ cm - sec}^{-1})}{(3310 \times 10^{-8} \text{ cm})} - (3 \times 10^{-12} \text{ erg})$$

$$= (6 \times 10^{-12} \text{ erg}) - (3 \times 10^{-12} \text{ erg}) = 3 \times 10^{-12} \text{ erg}$$

$$= \frac{3 \times 10^{-12}}{1.6 \times 10^{-12}} = 1.875 \text{ eV}.$$

5. The wavelength of the photoelectric threshold of a metal is 2300 Å. Let's determine (i) the work function in eV and (ii) the maximum kinetic energy (in eV) of the photoelectrons ejected by ultraviolet light of wavelength 1800 Å. (h = 6.626 ×10^{-34} joule-sec).

Solution : (i) Threshold wavelength λ_0 = 2300Å = 2300 × 10^{-10} meter. If v_0 be the threshold frequency, then the work function of the surface is

$$W = hv_0 = \frac{hv}{\lambda_0}$$

$$= \frac{(6.626 \times 10^{-34} \text{ joule-sec}) \times (3 \times 10^{8} \text{ m-sec}^{-1})}{2300 \times 10^{-10} \text{ m}}$$

$$= 8.64 \times 10^{-19} \text{ joule}$$

$$= \frac{8.64 \times 10^{-19}}{1.6 \times 10^{-19}} = 5.4 \text{ eV}.[\because 1 \text{ eV} = 1.6 \times 10^{-19} \text{joule}]$$

(*ii*) Let K_{max} be the kinetic energy of the fastest photoelectrons when light of frequency v falls on the surface. Then, from the Einstein's photoelectric equation, we have

$$K_{max} = hv - W.$$

If λ be the incident wavelength, we have

$$K_{max} = \frac{hc}{\lambda} - W.$$

$$\text{Here}: \frac{hc}{\lambda} = \frac{(6.626 \times 10^{-34} \text{ joule-sec}) \times (3 \times 10^{8} \text{m-sec}^{-1})}{(1800 \times 10^{-10} \text{m})}$$

$$= 11.0 \times 10^{-19} \text{ joule} = \frac{11.0 \times 10^{-19}}{1.6 \times 10^{-19}} = 6.9 \text{ eV}.$$

and $W = 5.4$ eV (as obtained above).

$$\therefore \quad K_{max} = \frac{hc}{\lambda} - W = 6.9 - 5.4 - 1.5 \text{ eV}.$$

6. The threshold wavelength for photoelectric emission from a metallic surface is 5800Å. Let's find the maximum energy of the photoelectrons if the wavelength of the incident light is 4500Å. (h = 66 × 10^{-34} joule-sec, c = 3 × 10^8 m/sec).

The answer is 0.62 eV.

7. The threshold frequency for photoelectric emission in copper is 1.1 × 10^{15} sec^{-1}. The maximum energy of the photoelectrons when light of frequency 15 × 10^{16} sec^{-1} falls on copper surface may be foundout. Let's also calculate the retarding potential. (h = 6.62 × 10^{-34} joule-sec)

Solution : If v_0 be the threshold frequency, then according to Einstein's equation, the maximum kinetic energy of the photoelectrons is

$$K_{max} = hv - hv_0$$

$$= 6.62 \times 10^{-34} (1.5 \times 10^{15} - 1.1 \times 10^{15})$$

$$= 2.648 \times 10^{-19} \text{ joule}.$$

But 1 eV = 1.6×10^{-19} joule.

$$\therefore \quad K_{max} = \frac{2.648 \times 10^{-19}}{1.6 \times 10^{-19}} = 1.655 \text{ eV}.$$

The retarding potential V_0 is the potential which just stops the photoelectrons of maximum kinetic energy. Thus

$$V_0 = 1.655 \text{ volts}.$$

8. The work function of a metal surface is 12 eV. The kinetic energy of the fastest and slowest photoelectrons and the retarding potential when light of frequency 5.5 × 10^{14} sec^{-1} falls on the surface is calculated below. (h = 662 × 10^{-34} joule-sec).

Solution : Let K_{max} be the kinetic energy of the fastest photoelectrons. Then the photoelectric equation is

$$K_{max} = h\upsilon - W, \qquad \text{... } (i)$$

where W is the work function and υ the incident frequency. Let us evaluate $h\upsilon$ in electron-volts. This is

$$h\upsilon = (6.62 \times 10^{-34} \text{ joule-sec})$$
$$(5.5 \times 10^{14} \text{ sec}^{-1})$$
$$= 3.64 \times 10^{-19} \text{ joule.}$$

But $\quad 1 \text{ eV} = 1.60 \times 10^{-19}$ joule.

$$\therefore \quad h\upsilon = \frac{6.34 \times 10^{-19}}{1.60 \times 10^{-19}} = 2.27 \text{ eV.}$$

Substituting this value of $h\upsilon$ and the given value of W (= 1.2 eV) in eq (i), we get

$$K_{max} = 2.27 - 1.2 = 1.07 \text{ eV.}$$

The kinetic energy of the slowest photoelectrons is zero.

The stopping potential V_0 is measured by the kinetic energy K_{max} of the fastest photoelectrons which is 1.07 electron-volts.

$$\therefore \quad V_0 = 1.07 \text{ volts.}$$

9. ***The stopping potential for photoelectrons emitted from a surface illuminated by light of wavelength 5893 Å is 0.36 volt. Let's calculate the maximum kinetic energy of photoelectrons, the work function of the surface and the threshold frequency. (h = 6.62 × 10^{-34} joule-sec, c = 3.0 × 10^{8} m/sec and 1 eV = 1.6 × 10^{-19} joule.)***

Solution : If V_0 volt be the stopping potential, then the (maximum) kinetic energy of the fastest electrons is given by

$$K_{max} = V_0 \text{ electron-volt} = 0.36 \text{ eV.}$$

Let W be the work function of the surface illuminated by

light of frequency v (or wavelength λ). Then from the Einstein's photoelectric equation, we have

$$K_{max} = hv - W = \frac{hc}{\lambda} - W,$$

$$\therefore \qquad W = \frac{hc}{\lambda} - K_{max}$$

Substituting the given values, we have

$$W = \frac{(6.62 \times 10^{-34} \text{ joule-sec}) \times (3 \times 10^{8} \text{ m/sec})}{5893 \times 10^{-10} \text{ m}}$$

$$-(0.36 \times 1.6 \times 10^{-19} \text{ Joule})$$

$$= 3.37 \times 10^{-19} \text{ joule} - 0.576 \times 10^{-19} \text{ joule}$$

$$= 2.794 \times 10^{-19} \text{ joule}$$

$$= \frac{2.794 \times 10^{-19}}{1.6 \times 10^{-19}} = 1.746 \text{ eV}.$$

The threshold frequency is

$$v_0 = \frac{W}{h} = \frac{2.794 \times 10^{-19} \text{ joule}}{6.62 \times 10^{-34} \text{ joule - sec}}$$

$$= 4.22 \times 10^{14} \text{ hertz}.$$

10. *A radiation of frequency 10^{16} Hz falls on a photocathode and ejects electrons with maximum energy of 4.2 × $l0^{-19}$ J. If the frequency of radiation is changed to 5 × 10^{14} Hz, the maximum energy of ejected electrons becomes 09 x 10^{-19} J. Planck's constant, threshold frequency and work function of photo-cathode material is calculated below.*

Solution : The photoelectric equation is

$$K_{max} = hv - W,$$

where v is the incident frequency and W is the work function.

Substituting the given values in the two cases, we have

$$4.2 \times 10^{-19} \text{ J} = h\ (10 \times 10^{14} \text{ s}^{-1}) - W \qquad \text{... } (i)$$

and $$0.9 \times 10^{-19} \text{ J} = h\ (5 \times 10^{14} \text{ s}^{-1}) - W. \qquad \text{... } (ii)$$

Subtracting eq. (*ii*) from eq. (*i*), we get

$$3.3 \times 10^{-19} \text{ J} = h\ (5 \times 10^{14} \text{ s}^{-1})$$

or $$h = \frac{3.3 \times 10^{-19} \text{ J}}{5 \times 10^{14} \text{ s}^{-1}} = 6.6 \times 10^{-34} \text{ J-s}$$

Putting this value of h in eq. (*i*) and solving, we get

$$W = 2.4 \times 10^{-19} \text{ J}.$$

The threshold frequency is

$$v_0 = \frac{W}{h} = \frac{2.4 \times 10^{-19} \text{ J}}{6.6 \times 10^{-34} \text{ J-s}} = 3.6 \times 10^{14} \text{ s}^{-1}.$$

11. When tungsten surface is illuminated with light of wavelength 1800 Å, the maximum kinetic energy of the electrons liberated is 1.5 eV. Photoelectric emission ceases when the wavelength of the incident light exceeds 2300Å. As for this case Planck's constant is calculated below ($c = 3.0 \times 10^8$ m/s, 1 eV = 1.6×10^{-19} J).

The maximum kinetic energy of photoelectrons emitted from a surface of work function W is given by

$$K_{max} = hv - W, \qquad \text{... } (i)$$

where v is the frequency of incident light.

For $$\lambda = 1800 \text{ Å},\ v = \frac{c}{\lambda} = \frac{3.0 \times 10^8 \text{ m/s}}{1800 \times 10^{-10} \text{ m}}$$

$$= 1.66 \times 10^{15} \text{ sec}^{-1}$$

and $$K_{max} = 1.5 \text{ eV} = 1.5 \times 1.6 \times 10^{-19}$$

$$= 2.4 \times 10^{-19} \text{ J}.$$

For $\lambda = 2300\text{Å}$,

$$v = \frac{3.0 \times 10^8 \text{ m/s}}{2300 \times 10^{-10} \text{ m}}$$

$$= 1.30 \times 10^{15} \text{ sec}^{-1}.$$

and $K_{max} = 0$

Substituting these values in eq. (*i*), we get

$$2.4 \times 10^{-19} \text{ J} = h(1.66 \times 10^{15} \text{ s}^{-1}) - W. \quad \text{... (ii)}$$

and $$0 = h\,(1.30 \times 10^{15} \text{ s}^{-1}) - W. \quad \text{... (iii)}$$

Subtracting eq. (*iii*) from eq (*ii*), we get

$$2.4 \times 10^{-19} \text{ J} = \text{h}\,(0.36 \times 10^{16} \text{ s}^{-1})$$

or $$h = \frac{2.4 \times 10^{-19} \text{ J}}{0.36 \times 10^{15} \text{ s}^{-1}}$$

$$= 6.66 \times 10^{-34} \text{ J-s}.$$

12. ***Light of wavelength 3000 Å falls on a metal surface having a work function of 2.3 eV. The maximum velocity of the ejected electrons is calculated as follows. (h = 6.6 × 10^{-34} joule-sec, c = 3.0 × 10^{8} m/sec, electronic mass m = 9.1 × 10^{-31} kg and 1 eV = 1.6 × 10^{-19} joule).***

Solution : The photoelectric equation is

$$K_{max} = hv - W = \frac{hc}{\lambda} - W$$

$$= \frac{(6.6 \times 10^{-34} \text{ joule sec})\,(3.0 \times 10^{8} \text{ m sec}^{-1})}{3000 \times 10^{-10} \text{ m}}$$

$$-(2.3 \times 1.6 \times 10^{-19} \text{ joule})$$

$$= 6.6 \times 10^{-19} \text{ joule} - 3.7 \times 10^{-19} \text{ joule}$$

$$= 2.9 \times 10^{-19} \text{ joule}.$$

Now, $K_{max} = \frac{1}{2} mv_{max}^2$

$$\therefore \quad v_{max} = \sqrt{\left(\frac{2K_{max}}{m}\right)}$$

$$= \sqrt{\left(\frac{2 \times 2.9 \times 10^{-19}}{9.1 \times 10^{-31}}\right)} = 8 \times 10^5 \text{ m/sec.}$$

13. The work function of a metal is 2.30 eV. The maximum wavelength of light which can liberate photo electrons from it, and the maximum velocity of the photoelectrons emitted by light of wavelength 2537 Å may be calculated as follows. ($h = 6.62 \times 10^{-27}$ erg-sec, $c = 3.0 \times 10^{10}$ cm/sec, electronic mass $m = 9.1 \times 10^{-28}$ gm and 1 eV = 1.6×10^{-12} erg).

The answer is 5400 Å, 9.54×10^7 cm/sec.

14. A beam of X-rays is scattered by free electrons. At 45° from the beam direction the scattered X-rays have a wavelength of 0.022 Å. The wavelength of the X-rays in the direct beam is given below.

Solution : The Compton shift (increase in wavelength) is given by

$$\Delta\lambda = \frac{h}{m_0 c}(1 - \cos\theta)$$

$$= 0.0243\ (1 - \cos\theta)\ \text{Å}.$$

For θ = 45°, we have

$$\Delta\lambda = 0.0243\ (1 - 0.707)\,[\because \cos 45° = 0.707]$$

$$= 0.0071\ \text{Å}.$$

The scattered wavelength is λ = 0.022 Å. Therefore, the incident wavelength must be

$$\lambda = \lambda' - \Delta\lambda$$

$$= 0.022 - 0.0071 = 0.0149\ \text{Å}.$$

15. The maximum change in wavelength in a Compton scattering is calculated as follows.

Solution : The Compton change in wavelength is given by

$$\Delta \lambda = \frac{h}{m_0 c}(1 - \cos\theta)$$

$$= 0.0243\ (1 - \cos\theta)\ \text{Å}.$$

The change in wavelength is maximum when the photon recoils back, *i.e.* when $\theta - 180°$. Then we have

$$(\Delta\lambda)_{max} = 0.0243\ (1 - \cos 180°) = 0.0486\ \text{Å}.$$

16. X-rays of wavelength 1.0 Å are scattered at such an angle that the recoil electron has the maximum kinetic energy. The wavelength of the scattered ray and the energy of recoil electrons is calculated as follows. h = 6.63 × 10^{-34} J–s, c = 3 × 10^8 m/s.

Solution : The recoil electron acquires maximum kinetic energy when the Compton shift is maximum (or the X-rays are scattered at 180°).

The Compton shift (change in wavelength) is given by

$$\Delta \lambda = 0.0243\ (1 - \cos\theta)\ \text{Å}.$$

It is maximum when $\theta = 180°$ *i.e.* $\cos\theta = -1$. Thus

$$(\Delta \lambda)_{max} = 0.0486\text{Å}.$$

The incident wavelength is $\lambda = 1.0$Å. Therefore, the scattered wavelength will be

$$\lambda' = \lambda + \Delta \lambda$$

$$= 1.0 + 0.0486 = 1.0486\ \text{Å}.$$

The energy of the incident X-ray photon is $h\nu$ ($= hc/\lambda$) and that of the scattered photon is $h\nu'$ ($= hc/\lambda'$). The balance has been imparted to the recoiling electron. Let it be K. Then

$$K = \frac{hc}{\lambda} - \frac{hc}{\lambda'}$$

$$= \frac{hc\left(\lambda' - \lambda\right)}{\lambda\lambda'} = \frac{hc\,\Delta\lambda}{\lambda\lambda'}.$$

Substituting the values of h, c, λ, λ' and $(\Delta\lambda)_{max}$, the maximum kinetic energy is

$$K = (6.63 \times 10^{-34} \text{ joule sec})$$

$$\times (3 \times 10^{8} \text{ meter-sec}^{-1})$$

$$\times \frac{\left(0.0486 \times 10^{-10} \text{ meter}\right)}{\left(1.0 \times 10^{-10} \text{ meter}\right) \times \left(1.0486 \times 10^{-10} \text{ meter}\right)}$$

$$= 0.92 \times 10^{-16} \text{ joule}$$

$$= \frac{0.92 \times 10^{-16}}{1.6 \times 10^{-19}} = 575 \text{ eV}.$$

17. X-rays of λ = 1.0 Å are scattered at 60° from the free electrons. After calculating the shift in wavelength and the energy of the recoil electrons. The answer is 0.012Å, 147 eV.

An X-ray photon is found to have its wavelength doubled on being scattered through 90°. The wavelength and energy of the incident photon will be as follows. (m_0 of electron = 9.0 × 10^{-28} gm).

Solution : The Compton change in wavelength is given by

$$\Delta\lambda = \frac{h}{m_0 c}\left(1 - \cos\theta\right).$$

When θ = 90°, $\Delta\lambda = \lambda$ (wavelength is doubled). Thus

$$\lambda = \frac{h}{m_0 c} = 0.0243 \text{ Å}$$

This is the wavelength of the incident photon. Its energy would be

$$E = h\upsilon = \frac{hc}{\lambda}$$

$$= \frac{hc}{h / m_0 c} = m_0 c^2$$

$$= (9.0 \times 10^{-28} \text{ gm}) \times (3.0 \times 10^{-28} \text{ cm/sec})^2$$

$$= 8.1 \times 10^{-7} \text{ erg.}$$

18. It is interested to find answer as to for what wavelength photon does Compton scattering result in a photon whose energy is one-half that of the original, at a scattering angle of 45° and in which region of electromagnetic spectrum does such a photon lie (Compton wavelength of the electron = 0.0242Å). Solution is found as follows

The Compton change in wavelength at 45° scattering is given by

$$\Delta\lambda = \frac{h}{m_0 c}\left(1 - \cos 45°\right)$$

$$= 0.0242\text{Å } (1 - 0.707)$$

$$= 0.0071 \text{ Å.}$$

The energy of the incident photon is hc/λ and that of the scattered photon is $hc/(\lambda + \Delta\lambda)$. Here it is given that

$$\frac{hc}{\lambda + \Delta\lambda} = \frac{1}{2}\frac{hc}{\lambda}$$

or $\lambda = \Delta\lambda = 0.0071 \text{ Å}$

This wavelength lies in the y-region of the electromagnetic spectrum.

19. X-rays of wavelength 1.0 Å are scattered from a carbon block. Let's find (i) the wavelength of the scattered beam in a direction making 90° with the incident beam, (ii) kinetic energy imparted to the recoiling electron, (iii) direction of the recoiling electron, ($h = 6.63 \times 10^{-34}$ joule-sec, $c = 3.0 \times 10^8$ meters/sec and 1 eV = 1.6×10^{-19} joule).

Solution : (*i*) The Compton shift is

$$\Delta \lambda = 0.0243(1\text{-cose } \theta) \text{ Å}.$$

At $\theta = 90°$, we have

$$\Delta \lambda = 0.0243\ (1 - \cos 90°) = 0.0243\text{Å}$$

Hence the wavelength of the scattered beam is

$$\lambda' = \lambda + \Delta \lambda = 1.0 + 0.0243 = 1.0243\text{Å}$$

(*ii*) The energy of the incident *X*-ray photon is $h\nu$ ($= hc/\lambda$) and that of the scattered photon is $h\nu'$ ($= hc/\lambda'$). The balance has been imparted to the recoiling electron. Let it be *K*. Then

$$K = \frac{hc}{\lambda} - \frac{hc}{\lambda'}$$

$$= \frac{hc\left(\lambda' - \lambda\right)}{\lambda\lambda'} = \frac{hc\ \Delta\lambda}{\lambda\lambda'}$$

Substituting the values of h, c, λ, λ' and $\Delta\lambda$, we get

$$K = (6.63 \times 10^{-34} \text{ joule-sec})$$

$$\times (3 \times 10^{8} \text{ meter-sec}^{-1})$$

$$\times \frac{\left(0.0243 \times 10^{-10} \text{ meter}\right)}{\left(1.0 \times 10^{-10} \text{ meter}\right) \times \left(1.0243 \times 10^{-10} \text{ meter}\right)}$$

$$= 4.72 \times 10^{-17} \text{ joule}$$

$$= \frac{4.72 \times 10^{-17}}{1.6 \times 10^{-19}} = 295 \text{ eV}.$$

(*iii*) Suppose the recoiling electron appears in a direction making an angle ϕ with the direction of the incident photon.

Applying conservation of momentum along and perpendicular to the direction of the incident photon, we get

$$\frac{h\nu}{c} = \frac{h\nu'}{c} \cos 90° + m\nu \cos \phi$$

and $$0 = \frac{hv'}{c} \sin 90^{\circ} - mv \sin \phi.$$

Dividing, we get

$$\tan \phi = \frac{v'}{v} = \frac{\lambda}{\lambda'}$$

Substituting the values of λ and λ, we get

$$\tan \phi = \frac{1.0}{1.0243} = 0.98$$

$$\therefore \quad \phi = \tan{-1}\ (0.98) = 44^{\circ}.$$

20. *Let's in Compton effect, if the incident photon has wavelength 2.0×10^{-8} cm and $\theta = 90^{\circ}$ deduce (i) the wavelength of the scattered photon, (ii) the energy of the recoil electron, (ii) the angle at which the recoil electron appears. (A= 6.62 $\times 10^{-27}$ erg-sec, c = 3.0×10^{10} cm-sec^{-1}, 1 eV = 1.6×10^{-12} erg).*

Answer is (i) 2.0243Å, (ii) 74.5 eV, (iii) $\tan^{-1}$ (0.99).

21. *In Compton scattering the incident radiation has a wavelength of 2A, and that scattered through 180° has 2.048 Å. The energy of recoil electron which scatters radiation through 60°. (Values of A, c and 1 eV same as above) may be calculated.*

The answer is 37 eV.

22. *Photons of wavelength 0.0124Å are scattered by free electrons at angles 90° and 180'. The wavelengths of the scattered photons and the energies transferred to the free electrons in the two cases found out as follows (A = 6.62×10^{-34} joule-sec, c - 3.0×10^{8} m/sec, $m_0 = 9.1 \times 10^{-31}$ kg and 1 eV = 1.6×10^{-19} joule).*

Solution : The Compton-increase in wavelength is

$$\Delta\lambda = \frac{h}{m_0 c} (1 - \cos \theta) = 0.0243\ (1 - \cos \theta)\ \text{Å}$$

For $\theta = 90^{\circ}$;

$$\Delta \lambda = 0.0243\ (1 - \cos 90^{\circ}) = 0.0243\ \text{Å}.$$

Therefore, the scattered wavelength is

$$\lambda' = \lambda + \Delta \lambda = 0.0124 + 0.0243 = 0.0367 \text{ Å}.$$

The kinetic energy transferred to the free electron is

$$K = \frac{hc\,\Delta\lambda}{\lambda\lambda'}$$

$$= \frac{(6.62 \times 10^{-34} \text{ joule-sec})(3.0 \times 10^{8} \text{ m/sec})(0.0243 \times 10^{-10} \text{ m})}{(0.0124 \times 10^{10} \text{ m})(0.0367 \times 10^{-10} \text{ m})}$$

$$= 1.06 \times 10^{-13} \text{ joule}$$

$$= \frac{1.06 \times 10^{-13}}{1.6 \times 10^{-19}} = 0.66 \times 10^{6} \text{ eV} = 0.66 \text{ MeV}.$$

For $\theta = 180°$; we have

$$\Delta \lambda = 0.0243\,(1 - \cos 180°) = 0.0486 \text{ Å}.$$

Therefore, the scattered wavelength is

$$\lambda' = \lambda + \Delta \lambda = 0.0124 + 0.0486 = 0.0610 \text{ Å}.$$

The kinetic energy transferred to the free electron is

$$K = \frac{hc\,\Delta\lambda}{\lambda\lambda'}$$

$$= 1.28 \times 10^{-13} \text{ joule} \qquad \text{(as calculated above)}$$

$$= 0.80 \text{ MeV}.$$

24. A beam of y-radiation having photon energy of 510 keV is incident on a foil of aluminium. The wavelength of the radiation scattered at 90° and also the energy and direction of emission of the recoil electron may be calculated.

Solution : The energy of the incident *y*-radiation is

$$E = 510 \text{ ke V} = 510 \times 10^{3} \text{ eV}$$

$$= 510 \times 10^{3} \times 1.6 \times 10^{-19}$$

$$= 8.16 \times 10^{-14} \text{ joule}.$$

Its wavelength is given by

$$\lambda = \frac{hc}{E}$$

$$= \frac{(6.62 \times 10^{-34} \text{ joule - sec}) \times (3.0 \times 10^{8} \text{ m / sec})}{8.16 \times 10^{-14} \text{ joule}}$$

$$= 2.43 \times 10^{-12} \text{ meter} = 0.0243\text{Å}.$$

The Compton-shift in wavelength at scattering angle of 90° is given by

$$\Delta \lambda = 0.0243 \text{ (l-cos } 90°) = 0.0243\text{Å}.$$

Hence the wavelength of the scattered radiation is

$$\lambda' = \lambda + \Delta \lambda = 0.0243 + 0.0243 = 0.0486 \text{ Å}.$$

The kinetic energy of the recoiling electron is given by

$$K = \frac{hc\,\Delta\lambda}{\lambda\lambda'}$$

$$= \frac{(6.62 \times 10^{-34} \text{ joule - sec})(3.0 \times 10^{8} \text{ m / sec})(0.0243 \times 10^{-1} \text{ m})}{(0.0243 \times 10^{-10}\text{m})(0.0486 \times 10^{-10}\text{m})}$$

$$= 4.1 \times 10^{-14} \text{ joule} = \frac{4.1 \times 10^{-14}}{1.6 \times 10^{-19}} = 2.56 \times 10^{5} \text{ eV}$$

$$= 256 \text{ KeV}$$

The direction ϕ of the recoil electron *for 90° scattering* is given by

$$\tan \phi = \frac{v'}{v} = \frac{\lambda}{\lambda'} = \frac{0.0243}{0.0486} = 0.5$$

$$\therefore \phi = \tan^{-1} (0.5) = 26.5°.$$

25. y-rays of energy 0.88 MeV are made to fall on a sheet of aluminium. Calculate the maximum energy of Compton recoil electrons is calculated as follows. ($h = 6.62 \times 10^{-34}$ joule-sec, $c = 30 \times 10^{8}$ m/sec and 1 eV = 1.6×10^{-18} joule).

Solution : The wavelength of the y-rays of energy 0.88 MeV is given by

$$\lambda = \frac{hc}{E}$$

$$= \frac{(6.62 \times 10^{-34} \text{ joule-sec})(3.0 \times 10^{8} \text{ m/sec})}{(0.88 \times 10^{6} \times 1.6 \times 10^{-19} \text{ joule})}$$

$$= 14.1 \times 10^{-13} \text{ m} = 0.0141 \text{ Å}.$$

The Compton wavelength shift is given by

$$\Delta\lambda = 0.0243\ (1 - \cos\theta)\ \text{Å}.$$

It is maximum for θ — 180°, and is given by

$$(\Delta\lambda)_{max} = 0.0486 \text{ Å}.$$

The corresponding scattered wavelength is

$$\lambda' = 0.0141 + 0.0486 = 0\ 0627 \text{ Å}.$$

The energy of recoil electrons is maximum for maximum Compton shift and is given by

$$K_{max} = \frac{hc(\Delta\lambda)_{max}}{\lambda\lambda'}$$

$$= \frac{(6.62 \times 10^{-34} \text{ joule-sec})(3.0 \times 10^{8} \text{ m/sec})(0.0486 \times 10^{-10} \text{ m})}{(0.0141 \times 10^{-10} \text{ m})(0.0627 \times 10^{-10} \text{ m})}$$

$$= 1.092 \times 10^{-13} \text{ joule}$$

$$= \frac{1.092 \times 10^{-13}}{1.6 \times 10^{-19}}$$

$$= 0.682 \times 10^{6} \text{ eV} = 0.682 \text{ MeV}$$

26. A monochromatic X-ray beam strikes a metal and a registers the photons. The adjacent detector scattered at 90°. The adjacent figure shows the energy spectrum of the photons observed at 90°. The parts which area modified due to Compton scattering is explained below.

Solution : The energy of the scattered photon is less than the energy of the incident photon. Hence the first peak, which corresponds to reduced energy is modified due to Compton scattering.

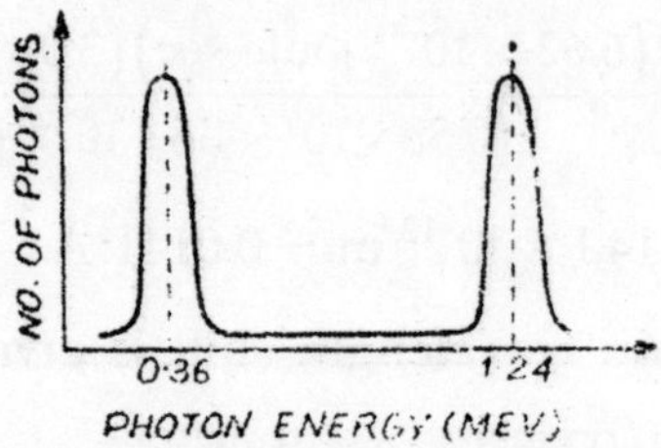

It can be worked out (as in above problems) that the wavelength of the 1.24-MeV incident photon is 0 01 Å, the Compton shift at 90° is 0.0243 Å, so that the modified wavelength is 0.0343Å. The energy corresponding to this wavelength is 0.36 MeV.

27. A photon collides with a free electron and is scattered at right angles to its initial direction. Let's find the percentage change in its energy as a result of this Compton collision, when the photon is a (i) microwave photon of λ = 3.0 cm, (ii) visible light photon of λ = 5000 Å (iii) X-ray photon of λ = 1.0 Å and (iv) y-ray photon of λ = 0.0124Å. The relative importance of the Compton effect in the different regions of electromagnetic spectrum is discussed below.

Solution : The change in photon energy is

$$\Delta E = hv—hv'$$

$$= \frac{hc}{\lambda} - \frac{hc}{\lambda'} = \frac{hc(\lambda' - \lambda)}{\lambda\lambda'} = \frac{hc\,\Delta\lambda}{\lambda\lambda'}$$

where $\Delta\lambda$ is the change (increase) in wavelength. Thus

$$\Delta E = \frac{hc\,\Delta\lambda}{\lambda(\lambda + \Delta\lambda)} \qquad [\therefore \lambda' - \lambda = \Delta\lambda]$$

The initial energy of the photon is $E = hc/\lambda$. Therefore, the fractional change in energy is

$$\frac{\Delta E}{E} = \frac{\Delta \lambda}{\lambda + \Delta \lambda}$$

Now, the Compton shift is given by

$$\Delta \lambda = 0.0243 \, (1 - \cos \theta) \, \text{Å}$$
$$= 0.0243\text{Å for } \theta = 90°.$$

$$\therefore \quad \frac{\Delta E}{E} = \frac{0.0243}{\lambda + 0.0243}$$

Let us now calculate $\frac{\Delta E}{E}$ for the different photons :

(*i*) For microwave photon (λ = 3.0 cm = 3.0 × 10^8 Å); we have

$$\frac{\Delta E}{E} = \frac{0.0243}{\left(3.0 \times 10^8\right) + 0.0243}$$
$$= 8 \times 10^{-11} = 8 \times 10^{-9} \, \%.$$

(*ii*) For visible light photon (λ = 5000Å), we have

$$\frac{\Delta E}{E} = \frac{0.0243}{5000 + 0.0243} = 4.8 \times 10^{-6} = 4.8 \times 10^{-4}\%$$

(*iii*) For the *X*-ray photon (λ = 1.0 Å), we have

$$\frac{\Delta E}{E} = \frac{0.0243}{1.0 + 0.0243} = 0.024 = 2.4\%$$

(*iv*) For the γ-ray photon (λ = 0.0124 Å), we have

$$\frac{\Delta E}{E} = \frac{0.0243}{0.0124 + 0.0243} = 0.66 = 66\%.$$

Thus the Compton effect is dominant only for the γ-ray region and the shorter *X*-ray region. It is not observable in the visible and microwave regions.

28. *The change in wavelength of a photon for 90° scattering from a neutron is calculated as follow (h = 6.6 × 10^{-34} J-s, c = 3 × 10^{8} m/s, neutron mass = 1.0087 amu and 1 amu = 1.66 × 10^{-27} kg.)*

Solution : The Compton change in wavelength is given by

$$\Delta\lambda = \frac{h}{m_0 c}(1 - \cos\theta)$$

where θ is the angle of scattering and m_0 is the rest mass of the scattering particle (here neutron). For θ = 90° we have

$$\Delta\lambda = \frac{h}{m_0 c}$$

$$= \frac{6.6 \times 10^{-34}\ \text{J-s}}{\left(1.0087 \times 1.66 \times 10^{-27}\ \text{kg}\right)\left(3 \times 10^{8}\ \text{m/s}\right)}$$

$$= 1.31 \times 10^{-15}\ \text{m}$$

$$= 1.31 \times 10^{-5}\ \text{Å}$$

9

Michelson-Morley Experiment

According to the wave theory of light, a light source sets up a disturbance travelling in all directions through a hypothetical medium called 'ether' which fills all space and penetrates all matter. The assumption of ether, however, created a problem. Does ether remain stationary in space when material bodies (including earth) move in it, or is it dragged along with the moving bodies ? Bradley' observation of the aberration of light from stars had indicated that *the ether must be stationary in space*. It means that if a material body, say earth, moves in space, there is a *relative* motion between the body and the ether. A number of experiments were performed to detect a relative motion between the earth and the ether. The most famous among them is the one performed by Michelson and Morley in 1887 using the Michelson interferometer.

A simplified plan of the experiment is shown in Fig. A beam of light from a source S falls upon a half-silvered glass plate P placed at 45° to the beam and is divided into two beams 1 and 2. The beams 1 and 2 travelling at right angles to each other, fall normally on mirrors M_1 and M_2 which reflect them back to *P*. The two beams returned to *P* are directed towards a telescope *T* in which interference fringes are observed.

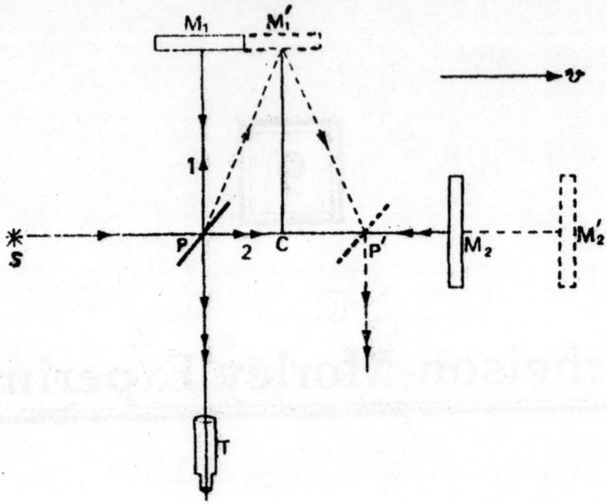

Let the mirrors M_1 and M_2 be at the same distance *I* from the plate *P*. Then if the apparatus were at rest in ether, the two beams would take the same time to return to *P*. But actually the earth, and hence the apparatus, is moving in space through the ether with a velocity v (say). Suppose this motion is in the direction of the initial beam of light. Then, if the initial beam strikes the plate *P* in the position shown, the paths of the two beams and the positions of their reflections from the mirrors will be as shown by the dotted lines. The time taken by the two beams on their journeys are not equal.

Let *c* be the velocity of light through the ether. The beam 2 moving towards M_2 has a velocity (c- v) *relative to the apparatus* on the outgoing trip, and *(c+v)* on the return trip. If t_2 be the total time taken by this beam to go from *P* to M_2 and back, then

$$t_2 = \frac{1}{c-v} + \frac{1}{c+v}$$

$$= \frac{2}{c^2 - c^2}$$

$$= \frac{2l}{c}\left(\frac{1}{1-\frac{v^2}{c^2}}\right). \qquad \text{... (i)}$$

The beam 1 moving *transversely* with respect to the apparatus retains its velocity c throughout. Let it take a time V to go from P to strike $M_{1,}$ travelling a distance ct'. In the same time, the mirror M_1 advances a distance vt'. Thus, in the right-angled triangle PM_1M_1', we have

$$PM_1'^2 = PM_1^2 + M_1M_1'^2.$$

But $\qquad PM_1 = I,\ M_1M_1' = vt'$ and $PM_1' = ct'$.

$$\therefore\ (ct')^2 = I^2 + (vt')^2$$

or $$t' = \frac{1}{\left(c^2 - v^2\right)^{1/2}} = \frac{1}{c}\frac{1}{\sqrt{\left(1-\frac{v^2}{c^2}\right)}}.$$

If t_1 be the total time taken by the beam to travel the whole path PM_1 'P', then

$$t_1 = 2t' = \frac{2l}{c}\frac{1}{\sqrt{\left(1-\frac{v^2}{c^2}\right)}}. \qquad \text{... (ii)}$$

Hence the difference between the times of travel of the two beams is, from (i) and (ii), given by

$$t_2 - t_1 = \frac{2l}{c}\left[\frac{1}{1-\frac{v^2}{c^2}} - \frac{1}{\sqrt{\left(1-\frac{v^2}{c^2}\right)}}\right].$$

$$= \frac{2l}{c}\left[\left(1-\frac{v^2}{c^2}\right)^{-1} - \left(1-\frac{v^2}{c^2}\right)^{-1/2}\right].$$

Using binomial expansion and dropping terms higher than the second order, we have

$$t_2 - t_1 = \frac{2l}{c}\left[\left(1+\frac{v^2}{c^2}\right)-\left(1+\frac{1}{2}\frac{v^2}{c^2}\right)\right]$$

$$= \frac{2l}{c}\left[\frac{1}{2}\frac{v^2}{c^2}\right] = \frac{lv^2}{c^2}.$$

The path difference δ between the beams corresponding to a time difference $(t_2—t_1)$ is

$$\delta = c\ (t_2 - t_1) = \frac{lv^2}{c^2}.$$

If the interferometer is suddenly brought to rest (v is made zero), then the path difference 5 would become zero. We know that if the path difference between two interfering waves changes by A, there is a shift of one fringe across the cross-wires in the field of view. Thus, if ΔN *is* the number of fringes which shift when the interferometer is stopped, then

$$\Delta N = \frac{\delta}{\lambda} = \frac{lv^2}{c^2\lambda}.$$

In the actual experiment, the whole apparatus, which was placed on a block of stone floated on mercury, was rotated through 90°. This introduced a path difference of the same amount in the opposite direction. Hence a shift of $\frac{2lv^2}{c^2\lambda}$ was expected.

To have an observable shift, Michelson and Morley increased the effective value of *I* upto 11 meters by reflecting the light back and forth several times. Then using values : l = 11 meter, $v = 3 \times 10^4$ meter/sec, $c = 3 \times 10^8$ meter/sec and $\lambda = 5.5 \times 10^{-7}$ meter (for visible light), the expected shift is

$$\Delta N = \frac{2lv^2}{c^2\lambda} = \frac{2 \times 11 \times \left(3\times 10^4\right)^2}{\left(3 \times 10^8\right)^2 \times 5.5 \times 10^{-7}} = 0.4,$$

or a shift of four-tenths a fringe.

Michelson and Morley were extremely surprised to see that there was *no* shift in the fringes when the interferometer was rotated through 90°. They repeated the experiment during various times of the day and various seasons of the year but no shift was observed. Trouton and Noble, in 1902, performed an electromagnetic experiment for the same purpose but with no positive result. Thus, *the motion of the earth through the ether could not be experimentally detected.*

Explanation of the Negative Result

Three separate explanations were given to the negative result of the Michelson-Morley experiment :

1. *Ether-drag Hypothesis : The* moving earth completely drags the ether with it so that there is no relative motion between the two and hence the question of shift does not arise. But this explanation was not accepted for two reasons : (i) It goes against the observed aberration of light from stars. (ii) Fizeau had experimentally shown that a moving body could drag the light waves only partially. Furthermore, this partial dragging of light waves was explained by the electromagnetic theory, without introducing the ether-drag hypothesis.

2. *Fitzgerald-Lorentz Contraction Hypothesis : Fitzgerald* and Lorentz independently put an adhoc hypothesis that all material bodies moving through the ether are contracted in the direction of motion by a factor $\sqrt{[1 - (v^2/c^2)]}$. It is easily seen that such a contraction in the interferometer arm would equalize the times t_1

and t_2 and no fringe-shift would be expected. This explanation also, being purely ad hoc, could not be accepted. Further, Rayleigh worked out that such a contraction is expected to produce double refraction which was however, never observed.

3. *Light Velocity Hypothesis :* The light from a moving source has a velocity which is the vector sum of its natural velocity and the velocity of the source. If this were true, then the light from the source had always the same constant velocity relative to the source and to the interferometer and the negative result could be explained. This explanation was also rejected because it was in conflict with the wave theory of light as well as with certain astronomical evidences concerning double stars.

Einstein's Revolutionary Idea

Einstein, in 1905, proposed a new revolutionary idea that the *motion through ether is a meaningless concept ; only motion relative to a frame of reference has physical significance.* The frame of reference may be a road, the earth's surface, the sun, the centre of our galaxy ; but in every case we must specify it. If we were isolated in the universe, there would be no way in which we could determine whether we are in motion or not. That is why it is impossible to perform any experiment for detecting earth's motion through ether. This idea was ultimately developed in the theory of relativity.

Einstein's Special Theory of Relativity : In 1905, Einstein published his special theory of relativity which is based upon two postulates.

Postulate I: The laws of Physics have the same form in all inertial frames of reference moving with a 'constant' velocity relative to one another (The Principle of Relativity). This postulate expresses the absence of a universal frame of reference. If the laws of Physics were different for observers in different frames in relative motion, it could be determined from these differences which objects are stationary in space and which are moving. But because there is no universal frame of reference, this distinction between objects cannot be made. (If we are in a free balloon from which we cannot see anything except another free balloon which is changing its position relative to us, we cannot know which balloon is really moving).

Postulate II: The speed of light in free space is the same in all inertial frames of reference (The Principle of the Constancy of the Speed of Light). This postulate follows directly from the result of the Michelson Morley experiment.

Galilean-Newtonian Transformation Equations

Let S and S' (Fig.) be two frames of reference, S' moving along the + x direction with a constant velocity v relative to S Let (x, v, z, t) be the space and time coordinates of an event for an observer in the frame S, and (x', y', z', t') the coordinates of the same event for an observer in the frame S'. Let the time be counted from the instant when the origins O and O' momentarily coincide. We have to see how are the measurements x, y, z, t related to x', y', z', t'.

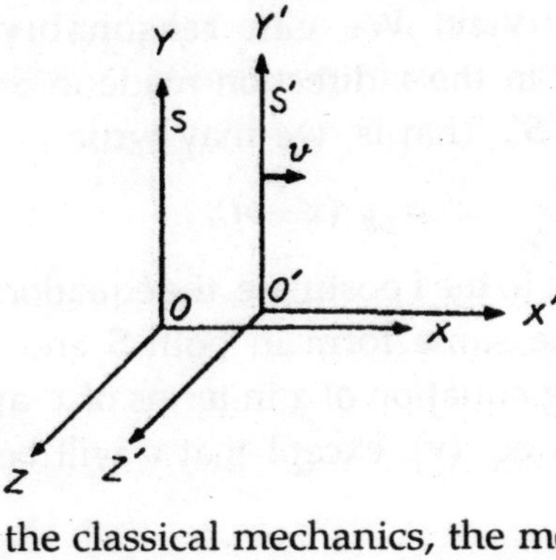

According to the classical mechanics, the measurement in the x direction made in S' will be smaller than that made in

S by the amount vt, which is the distance moved by S' in the x direction. That is

$$x' = x - vt. \quad \text{... (i)}$$

There is no relative motion in the y and z directions, and so

$$y' = y \quad \text{... (ii)}$$

and

$$z' = z. \quad \text{... (iii)}$$

In classical mechanics the time-intervals measured in different frames of reference in relative motion are the same, and so

$$t' = t. \quad \text{... (iv)}$$

The set of equations (i) to (iv) are the Galilean-Newtonian transformation equations. They, however, violate both of the postulates of special relativity. The equations of electromagnetism assume very different forms when these equations are used to convert quantities measured in one frame into the equivalent quantities in the other. This violates the I postulate. Again, if we measure the speed of light along the x direction in the frame S to be c, then in the frame S' it will be $c' = c - v$. This violates the II postulate.

Lorentz Transformation Equations : Let us now develop a set of transformation equations directly from the postulates of special relativity. We can reasonably assume that a measurement in the x direction made in S is proportional to that made in S'. That is, we may write

$$x' = y\,(x - vt), \quad \text{... (v)}$$

According to the I postulate, the equations (laws) of Physics must have the same form in both S and S'. Therefore, the corresponding equation of x in terms of x' and t' will be of the same form as eq. (v), except that v will be replaced by $-v$. Thus

$$x = y\,(x' + vt'). \quad \text{... (vi)}$$

The time coordinates t and t', however, are not equal. Let us substitute the value of x' from eq. (v) in eq. (vi) :

$$x = y\,[\,y\,(x - vt) + vt']$$

or
$$\frac{x}{y} = yx - yvt + vt'.$$

$\therefore$
$$t' = \frac{x}{yv} - \frac{yx}{v} + yt$$

$$= \text{yt} - \frac{yx}{v}\left(1 - \frac{1}{y^2}\right) \qquad \text{... (vii)}$$

Similarly, we can show that

$$t = yt' + \frac{yx'}{v}\left(1 - \frac{1}{y^2}\right). \qquad \text{...(viii)}$$

Now, y can be evaluated from the II postulate. Suppose a light signal is given at the origin O at time $t = 0, t' = 0$, *i.e.* when O and O' coincide. The signal travels with a speed c which is same for both the frames (II postulate). Its position as seen from S and S' after some time is given by

$$x = ct$$

and
$$x' = ct'.$$

Substituting for x and x' in eq. (v) and (vi), we obtain

$$ct' = \text{yt}\,(c - v)$$

and
$$ct = \text{yt}'\,(c + v).$$

Solving these two equations, we can easily see that

$$y = \frac{1}{\sqrt{\left[1 - \left(v^2/c^2\right)\right]}} \qquad \text{... (ix)}$$

Inserting this value of y in eq. (v) and (vi), we get

$$x' = \frac{x - vt}{\sqrt{\left[1-\left(v^2/c^2\right)\right]}} \qquad \text{... (x) a}$$

and
$$x = \frac{x' + vt'}{\sqrt{\left[1-\left(v^2/c^2\right)\right]}} \qquad \text{...(x) b}$$

Again, inserting the value of y, or $1 - \frac{1}{y^2} = \frac{v^2}{c^2}$, in eq. (vii) and (viii), we get

$$t' = yt - \frac{yx}{v}\left(\frac{v^2}{c^2}\right) = y\left(t - \frac{xv}{c^2}\right)$$

or
$$t' = \frac{t - \left(xv/c^2\right)}{\sqrt{\left[1-\left(v^2/c^2\right)\right]}} \qquad \text{... (xi) a}$$

and similarly
$$t = \frac{t' + \left(x'v/c^2\right)}{\sqrt{\left[1-\left(v^2/c^2\right)\right]}} \qquad \text{... (xi) b}$$

In the general case when the light signal is not restricted to travel along the *x*-axis, we get two more relations

and
$$\left.\begin{aligned} y &= y \\ z' &= z, \end{aligned}\right\} \qquad \text{... (xii)}$$

since these distances are measured perpendicular to *v*.

Eq. (x), (xi) and (xii) are known as Lorentz transformation equations. They are used for the transformation of the space and time coordinates of an event from one frame of reference into the other in uniform relative motion.

There are two important aspects of the Lorentz equations:

(i) The measurements of position and time depend upon the frame of reference of the observer. Two events which are simultaneous when viewed from one frame of reference are not simultaneous when viewed from a second frame which is moving relative to the first.

(ii) The Lorentz equations reduce to the classical Galilean equations when $v << c$. Therefore, the consequences of special relativity are not apparent unless enormous velocities are involved.

Invariancy of $x^2 + y^2 + z^2 - c^2t^2$: Let (x, y, z, t) and $(x'. y', z', t')$ be the space-time coordinates of an event observed by two observers in reference frame S and S' respectively, the frame S' moving with a constant velocity v relative to S. Using Lorentz transformation equations, we have to simply prove that

$$x^2 + y^2 + z^2 - c^2t^2 = x'^2 + y'^2 + z'^2 - c^2t'^2.$$

The Lorentz equations are

$$x' = \frac{x - vt}{\sqrt{\left[1-\left(v^2/c^2\right)\right]}},$$

$$y' = y \; z' = z, \; t' = \frac{x-\left(xv/c^2\right)}{\sqrt{\left[1-\left(v^2/c^2\right)\right]}}$$

Then we have

$$x'^2 + y'^2 + z'^2 - c^2t'^2$$

$$= \frac{(x-vt)^2}{1-\left(v^2/c^2\right)} + y^2 + z^2 - \frac{c^2\left\{t-\left(xv/c^2\right)\right\}^2}{1-\left(v^2/c^2\right)}$$

$$= \frac{1}{1-\left(v^2/c^2\right)}\left[x^2 + v^2t^2 - 2xvt - c^2\left(t^2 + \frac{x^2v^2}{c^4} - \frac{2t\,xv}{c^2}\right)\right] + y^2 + z^2$$

$$= \frac{1}{1-\left(v^2/c^2\right)}\left[x^2 + v^2t^2 - 2xvt - c^2t^2 - \frac{x^2v^2}{c^2} + 2txv\right] + y^2 + z^2$$

$$= \frac{1}{1-(v^2/c^2)}\left[x^2\left(1-\frac{v^2}{c^2}\right)-c^2t^2\left(1-\frac{v^2}{c^2}\right)\right]+y^2+z^2$$

$$= [x^2 - c^2t^2] + y^2 + z^2 = x^2 + y^2 + z^2 - c^2t^2.$$

Hence $x^2 + y^2 + z^2 - c^2t^2$ is Lorentzo-invariant.

Let us consider the given equation

$$x'^2 = c^2 t'^2.$$

By Lorentz transformation, we have

$$x' = \frac{x - vt}{\sqrt{[1-(v^2/c^2)]}} \text{ and } t' = \frac{t-(xv/c^2)}{\sqrt{[1-(v^2/c^2)]}}.$$

Then, from the given equation, we have

$$= \frac{(x-vt)^2}{1-\frac{v^2}{c^2}} = \frac{c^2\{t-(xv/c^2)\}^2}{1-\frac{v^2}{c^2}}$$

or $$(x - vt)^2 = c^2\left(t^2 + \frac{x^2v^2}{c^4} - \frac{2t\,xv}{c^2}\right)$$

or $$x^2 + v^2t^2 - 2xvt = c^2t^2 + \frac{x^2v^2}{c^2} - 2txv$$

or $$x^2 - c^2t^2 + v^2t^2 - \frac{x^2v^2}{c^2} = 0$$

or $$(x^2 - c^2t^2) - \frac{v^2}{c^2}(x^2 - c^2t^2) = 0$$

or $$(x^2 - c^2t^2)\left(1 - \frac{v^2}{c^2}\right) = 0.$$

Since $v \neq c$, $$\left(1-\frac{v^2}{c^2}\right) \neq 0.$$

$\therefore$ $$x2 - c^2t^2 = 0$$

or $$x^2 = c^2t^2.$$

This result shows that the velocity of light is an absolute constant independent of the frame of reference.

Lorentz-Fitzgerald Length Contraction

Relative motion affects measurement of length. *A body moving with a velocity v relative to an observer appears to the observer to be contracted in length in the direction of motion by a factor* $\sqrt{\left[1-\left(v^2/c^2\right)\right]}$.

To prove this length contraction, let us consider a rod placed along the x'-axis of a moving frame of reference S'. Let x_1 and x_2' be the coordinates of the ends of the rod, as seen by an observer in S'. Then the length of the rod as observed from 5' is

$$L_0 = x_2' - x_1'.$$

L_0 is the rod's length as measured in a frame of reference in which it is at rest. Now, suppose another observer measures the length of the rod from a stationary frame of reference S, relative to which the rod (or the frame S') is moving with the velocity v. If he records the coordinates of the ends of the rod as x_1 and x_2 at a time t, the rod appears to him as having a length

$$L = x_3 - x_1.$$

From the Lorentz transformation for a given value of t, we have

$$x_2' = \frac{x_2 - vt}{\sqrt{\left[1-\left(v^2/c^2\right)\right]}}$$

and

$$x_1' = \frac{x_1 - vt}{\sqrt{\left[1-\left(v^2/c^2\right)\right]}}.$$

$$\therefore \qquad L_0 = x_2' - x_1' = \frac{x_2 - x_1}{\sqrt{\left[1-\left(v^2/c^2\right)\right]}}$$

$$= \frac{L}{\sqrt{\left[1-\left(v^2/c^2\right)\right]}}$$

or
$$L = L_0 \sqrt{\left[1-\left(v^2/c^2\right)\right]}.$$

The equation shows that to the stationary observer in S, the rod placed in the moving frame 5' appears to be contracted by a factor $\sqrt{\left[1-\left(v^2/c^2\right)\right]}$ This contraction occurs in the direction of relative motion. If v is parallel to x, the y and z dimensions of the rod are same in both S and S'.

If v — c, then $L = 0$. It means that a rod moving with the speed of light will appear as reduced to a point to a stationary observer.

Proper Length : The length L_0 measured in the frame in which the rod is at rest is called 'proper' length,

Time Dilation. Time-intervals are also affected by relative motion. *A clock moving with velocity v with, respect to an observer appears him to have slowed down by a factor* $1/\sqrt{\left[1-\left(v^2/c^2\right)\right]}$, *than when at rest with respect to him.*

Suppose a clock is placed at the point x' in the moving frame S'. An observer in S' finds that the clock gives two ticks at times t_1' and t_2' . The time-interval between the ticks as judged from S' is

$$t_0 = t_2' — t_1'.$$

t_0 is the interval as measured in a frame in which the clock is at rest. Now, suppose another observer measures the time-interval (with his own clock) between the same two ticks from a stationary frame of reference S, relative to which the clock (or S') is moving with velocity v. If he records the ticks at times t_1 and t_2, the time-interval appears to him as

$$t = t_2 - t_1.$$

From the Lorentz transformation for a fixed value of x', we have

$$t_2 = \frac{t_2' + (x'v/c^2)}{\sqrt{[1-(v^2/c^2)]}} \text{ and } t_1 = \frac{t_1' + (x'v/c^2)}{\sqrt{[1-(v^2/c^2)]}}$$

$$\therefore \quad t = t_2 - t_1 = \frac{t_2' - t_1'}{\sqrt{[1-(v^2/c^2)]}}$$

or

$$t = \frac{t_0}{\sqrt{[1-(v^2/c^2)]}}$$

This eq. shows that to the stationary observer in S, the time-interval appears to be *lengthened* by a factor $\frac{1}{\sqrt{[1-(v^2/c^2)]}}$

a stationary clock measures a longer time-interval between events occurring in a moving frame of reference than does a clock in the moving frame. In other words, a moving clock appear to be slowed down to a stationary observer. This phenomenon is known as "time dilation" as "time dilation.

If $v = c$, then $t = \infty$. It means that a clock moving with the speed of light will appear to be completely stopped to a stationary observer.

Proper Time-interval : The proper time-interval is the time-interval measured by a clock attached to the observed body. In the above example, t_0 is the proper time-interval, (t is non-proper time-interval).

Experimental Verification of Time Dilation : μ-Meson Decay: Direct Experimental evidence for time-dilation; has been found in cosmic ray Physics. Certain particles, called μ-mesons, are created by fast cosmic-ray particles at a height of about 10 kilometers from the surface of the earth and reach the earth in large numbers. They have a typical speed of 2.994×10^8 meters/sec, which is 0.998 of the speed of light *e*. Further, "a μ-meson is found to have an average lifetime of 2×10^{-6} second after which it decays into an electron. Thus, in its lifetime a μ-meson can travel a distance of only (2.994×10^8 m/sec) × (2×10^6 sec) ~ 600 meters or 0.6 km. Then how do μ-mesons travel a distance of H) km to reach the earth ?

The answer is supplied by the special theory of relativity. The μ-meson has a lifetime *to* (= 2×10^{-6} sec) *in its own frame of reference.* In the observer's frame of reference on the earth, however, the lifetime is lengthened owing to the relative motion, to the value t given by

$$t = \frac{t_0}{\sqrt{\left[1-\left(v^2/c^2\right)\right]}} = \frac{2 \times 10^{-6}}{\sqrt{[1-(0.998)^2]}}$$

$$= \frac{2 \times 10^{-6}}{0.063} = 3.17 \times 10^{-5} \text{ sec.}$$

In this lifetime a meson whose speed is 0.998 c (= 2.994×10^8 meters/sec) can travel a distance.

$$(2.994 \times 10^8 \text{ m/sec})\text{x}(3.17 \times 10^{-5} \text{ sec})$$

$$\simeq 9500 \text{ meters} \simeq 9.5 \text{ km.}$$

Hence, despite their brief lifetime, it is possible for the mesons to reach the earth from the large altitudes at which they are actually formed.

An interesting implication of the relativity theory is intervals are not the same for two observers in relative motion. This leads to an important fact that *two events which appear to take place simultaneously to one observer are, in general, not simultaneous to another observer in relative motion.*

Suppose two time-bombs explode at different places x_1 and x_2 but at the same time t_0 to an observer on the ground. The situation is different to a pilot of a spaceship moving with a velocity v relative to ground. To him the explosion at x_1 occurs at

$$t_1' = \frac{t_0 - \left(x_1 v / c^2\right)}{\sqrt{\left[1 - \left(v^2 / c^2\right)\right]}}$$

and that at x_2 occurs at

$$t_2' = \frac{t_0 - \left(x_2 v / c^2\right)}{\sqrt{\left[1 - \left(v^2 / c^2\right)\right]}}$$

Hence the two events (explosions) that occur simultaneously to one observer are separated to another observer by a time-interval of

$$t_2' - t_1' = \frac{\left(x_1 - x_2\right) v / c^2}{\sqrt{\left[1 - \left(v^2 / c^2\right)\right]}}.$$

Who is right ? The question is meaningless ; both observers are right because each one measures what he sees. Hence we conclude that there is no such thing as "absolute time" which is same for all observers. Time is relative and is different for observers in relative motion.

***Past, Present and Future* :** The relative character of time, however, does not discard all of our every-day experiences regarding time. For example, no observer, regardless of his state of motion, can see an event before it actually occurs. This is because an event sends out light signal which takes finite time to reach the observer. Similarly, time never runs backward to *any* observer. A sequence of events that occur somewhere at t_1 t_2, t_3.........will appear in the same order to all observers everywhere, though not necessarily with the same time-intervals $t_2 - t_1$, $t_3 - t_2$........between each pair of events. Thus we can order events in time. Hence there is a sense of time distinguishing past, present and future.

However, it is possible that an observer has seen an event occurring at a distant place, whereas a second farther observer has yet to see it. Thus the event has occurred in the past of the first observer, but has to occur in the future of the second observer. (In other words, the two observers disagree about past and future) Can the first observer send a radio message about this event to the second observer enabling him to predict the future ? The answer is "no" because the radio message has to travel through space with the speed of light and can never reach the second observer before he himself becomes aware of the event. Thus even in the theory of relativity there is no way to peep into the future.

Relativistic Addition of Velocities

Very high velocities cannot be added directly as in classical mechanics, because velocities must be added in a manner consistent with the Lorentz transformation.

Let there be two frames of reference S and S'', the frame S' moving with a constant velocity v relative to S along the x-axis. Let a body move a distance dx in a time-interval dt in the frame S. Then the velocity of the body measured by an observer in S is

$$u = \frac{dx}{dt}.$$

To an observer in S'', both the distance and the time-interval will appear different say, dx' and dt', so that to him the velocity of the body will be

$$u' = \frac{dx'}{dt'}.$$

From the Lorentz transformations, we have

$$x' = \frac{x - vt}{\sqrt{\left[1-\left(v^2 c^2\right)\right]}} \text{ and } t' = \frac{t-\left(xv/c^2\right)}{\sqrt{\left[1-\left(v^2 c^2\right)\right]}}$$

Differentiating, we get

$$dx' = \frac{dx - v\,dt}{\sqrt{\left[1-\left(v^2/c^2\right)\right]}} \text{ and } dt' = \frac{dt-\left(v\,dx/c^2\right)}{\sqrt{\left[1-\left(v^2/c^2\right)\right]}}$$

$$\therefore \qquad u' = \frac{dx'}{dt'} = \frac{dx - v\,dt}{dt - \dfrac{v\,dx}{c^2}} = \frac{\dfrac{dx}{dt} - v}{1 - \dfrac{v\,dx}{c^2\,dt}}.$$

But $dx/dt = u$.

$$\therefore \qquad u' = \frac{u - v}{1-\left(uv/c^2\right)}.$$

This is the *relativistic* addition of velocities u and v. In classical mechanics it is simply

$$u' = (u-v).$$

In the same way we may show that if the velocity u' with respect to frame S' is given, we can determine its value u with respect to S by the relation

$$u = \frac{u' + v}{1+\left(u'v/c^2\right)}.$$

There are two important consequences of the relativistic addition of velocities. If $u' = c$, *i e.,* if a ray of light is emitted in the moving frame S' along the x axis, an observer in frame S will measure the velocity as

$$u = \frac{u' + v}{1 + \left(u'v / c^2\right)} = \frac{c + v}{1 + \frac{cv}{c^2}} = \frac{c\left(c + v\right)}{c + v} = \text{c}.$$

Thus both observers measure the same value for the velocity of light. *The velocity of light in vacuum is thus an absolute constant independent of the frame of reference.* This is one of the postulates of relativity.

The above expression also shows that *any velocity v added relativistically to the velocity of light c merely reproduces* the *velocity of light.* It means that the velocity of light in vacuum is the maximum attainable velocity in nature.

Relativity of Mass

In the light of relativistic ideas of time and space and the way in which velocities add together, the concept of mass as an invariable property of body remaining constant under all circumstances has to be abandoned. In fact, mass is a function of the velocity of the body. It increases with increasing velocity according to the relation

$$m = \frac{m_0}{\sqrt{\left[1 - \left(v^2 / c^2\right)\right]}},$$

where m_0 is the 'rest mass' of the body (*i e.* the mass of the body when at rest), *c* is the velocity of light, and *v* the velocity of the body. This increase in mass has been directly verified in various electron diffraction experiments, and also in the operation of particle-accelerators.

The increase in mass is not appreciable until the velocity approaches that of light. It is, therefore, not ordinarily detected.

Derivation of the Formula: Let *S* and *S′* (Fig.) be two frames of reference, *S′* moving with a velocity *v* relative to *S*. Let two particles 1 and 2 of masses m_1 and m_2 and moving with velocities *u* and — *u′* in the frame *S″* approach each other.

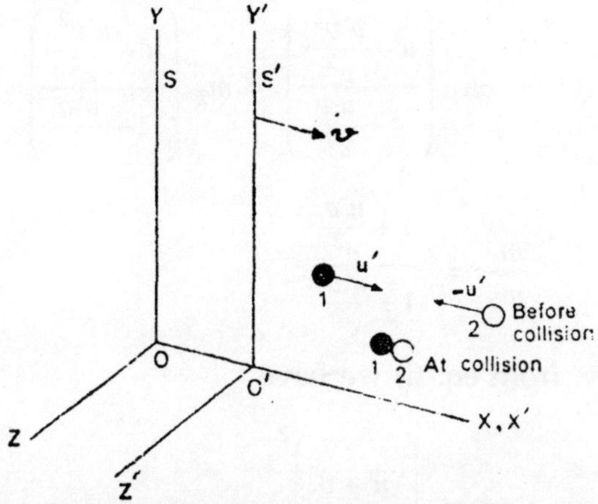

The velocities of the particles, as seen from the frame *S* will, however, be different and are given by the relativistic addition of velocities as

$$u_1 = \frac{u' + v}{1 + \frac{u'v'}{c^2}}, \qquad \text{...(i)}$$

and

$$u_2 = \frac{-u' + v}{1 - \frac{u'v'}{c^2}}. \qquad \text{... (iv)}$$

At the instant of collision, the two particles are momentarily at rest with respect to the frame *S'* but, as seen from the frame *S*, they are still moving with velocity v. Since the total momentum of the two particles is conserved, we have

$$m_1 u_1 + m_2 u_2 = (m_1 + m_2)\, v$$

or

$$m_1 \left(\frac{u' + v}{1 + \frac{u'v'}{c^2}} \right) + m_2 \left(\frac{-u' + v}{1 - \frac{u'v}{c^2}} \right) = (m_1 + m_2)\, v$$

or

$$m_1 \left(\frac{u' + v}{1 + \frac{u'v'}{c^2}} - v \right) = m_2 \left(v - \frac{-u' + v}{1 - \frac{u'v}{c^2}} \right)$$

or $$m_1 \left(\frac{u' - \frac{u'v^2}{c^2}}{1 + \frac{u'v}{c^2}} \right) = m_2 \left(\frac{u' - \frac{u'v^2}{c^2}}{1 - \frac{u'v}{c^2}} \right)$$

or $$\frac{m_1}{m_2} = \frac{1 + \frac{u'v}{c^2}}{1 - \frac{u'v'}{c^2}}. \quad \text{... (iii)}$$

Now, from eq. (i) we have

$$u_1^2 = \left(\frac{u' + v}{1 + \frac{u'v}{c^2}} \right)^2$$

or $$1 - \frac{u_1^2}{c^2} = 1 - \frac{1}{c^2} \left(\frac{u' + v}{1 + \frac{u'v}{c^2}} \right)$$

$$= \frac{\left(1 + \frac{u'v}{c^2}\right)^2 - \left(\frac{u' + v}{c}\right)^2}{\left(1 + \frac{u'v}{c^2}\right)^2} = \frac{\left(1 - \frac{u'^2}{c^2}\right)\left(1 - \frac{v^2}{c^2}\right)}{\left(1 + \frac{u'v}{c^2}\right)^2}$$

$$\therefore \quad 1 + \frac{u'v}{c^2} = \sqrt{\left[\frac{\left(1 - \frac{u'^2}{c^2}\right)\left(1 - \frac{v^2}{c^2}\right)}{1 - \frac{u_1^2}{c^2}}\right]}.$$

Similarly,

$$1 - \frac{u'v}{c^2} = \sqrt{\left[\frac{\left(1 - \frac{u'^2}{c^2}\right)\left(1 - \frac{v^2}{c^2}\right)}{1 - \frac{u_2^2}{c^2}}\right]}.$$

Substituting these values in eq. (iii), we have

$$\frac{m_1}{m_2} = \sqrt{\left[\frac{\left(1 - \frac{u_2^2}{c^2}\right)}{1 - \frac{u_1^2}{c^2}}\right]}.$$

Now if the particle 2 is moving with zero velocity in the frame S before collision, we have $u_2 = 0$ and $m_2 = m_0$ (rest mass of the particle). Then

$$m_1 = \frac{m_0}{\sqrt{\left(1 - \frac{u_1^2}{c^2}\right)}}.$$

Since the two particles are exactly identical, the rest mass of the other particle is also m_0. Hence the above equation can be considered as applying to *one and the same particle.* Then replacing m_1 by m and u_1 by v, the last equation becomes

$$m = \frac{m_0}{\sqrt{\left(1 - \frac{u^2}{c^2}\right)}}$$

This is the relativistic formula for the variation of mass with velocity. We can draw the following conclusions from it :

(*i*) As the velocity v of the particle relative to the observer increases, the mass of the particle increases.

(*ii*) As $v \to c$, $m \to \infty$. This means that *no material particle can have a velocity equal to, or greater than the velocity of light.*

(*iii*) When $v << c$, then v^2/c^2 can be ignored compared to 1, and we have $mc \simeq m_0$. This means that at ordinary velocities the difference between m and m_0 is too small to be detectable.

Experimental Verification: In 1909, Bucherer determined the ratio *e/m* as a function of velocity for 'fast' electrons emitted from radioactive nuclei and showed that the mass of the electron varies with its velocity according to the above relation.

Limiting Velocity of a Material Particle

The mass of a body moving relative to an observer increases with increasing velocity according to the equation

$$m = \frac{m_0}{\sqrt{\left[1-\left(v^2/c^2\right)\right]}},$$

where m is the mass at velocity v and m_0 the rest mass. Thus as the body's velocity v approaches the velocity of light c, the ratio $v^2/c^2 \to 1$ and the quantity $\sqrt{[1 \to (v^2/c^2)]} \to 0$. Hence the mass m of the body approaches infinity as v approaches c. For instance, an electron energised to 10^9 eV has an effective mass nearly 2000 times its rest mass.

The variation of mass with velocity, therefore, places a limit on the velocity which a material particle can have relative to an observer. As we accelerate the particle, its mass becomes larger and larger and it becomes increasingly more difficult to accelerate it further. In fact, since the mass would become infinite when $v = c$, an infinite force would be needed to accelerate the particle to the velocity of light c. Since no infinite forces are in the universe, the velocity of a particle can never be made to exceed the velocity of light in free space.

Although it is true that a moving particle can never travel faster than the velocity of light *in free space* ($c = 3 \times 10^8$ m/sec), this is not true in a material medium (water, glass, etc.). In all material media light travels more slowly than its velocity in free space. Very energetic atomic particles are, however, capable of moving faster in a material medium than the velocity of light *in that medium,* though never faster than the velocity of light in free space. This phenomenon gives rise to light waves known as 'cerenkov radiation.' Hence the motion of high-energy electron beam *across the screen* faster than light does *not* violate the theory of relativity.

Relativistic Momentum and Force

In relativity, the momentum p of a particle at any velocity v can be defined in the same way as in classical mechanics, that is,

$$p = mv.$$

But $m = m_0 / \sqrt{[1 - (v^2/c^2)]}$, where m_0 is the rest mass of the particle.

$$\therefore \qquad p = \frac{m_0 v}{\sqrt{\left[1-\left(v^2/c^2\right)\right]}}$$

With this definition of momentum, the law of conservation of momentum is true in special relativity just as in classical Physics.

In relativity, however, the force acting on a body cannot be defined as mass multiplied by acceleration of the body. It must be defined only as rate of change of momentum (Newton's original definition). Thus

$$F = \frac{dp}{dt} = \frac{d}{dt}\left[\frac{m_0 v}{\sqrt{\left[1-\left(v^2/c^2\right)\right]}}\right].$$

Rest Mass of Photon

A photon travels with the velocity of light, and hence we must use relativistic expression for its momentum, p. Thus, for a photon

$$p = \frac{m_0 v}{\sqrt{\left[1-\left(v^2/c^2\right)\right]}}$$

where m_0 is the rest mass and v the velocity of the photon.

From quantum theory of radiation, the momentum of a photon of radiation of wavelength λ is $p = h/\lambda$ where h is the Planck's constant. Therefore

$$\frac{h}{\lambda} = \frac{m_0 v}{\sqrt{\left[1-\left(v^2/c^2\right)\right]}}$$

or $$m_0 = \frac{h}{v\lambda}\sqrt{\left[1-\left(v^2/c^2\right)\right]}.$$

Since for photon, $v = c$, we have

$$m_0 = 0.$$

Hence the rest mass of photon is zero.

Einstein's Mass-Energy Equivalence Relation

According to Einstein's special theory of relativity, the mass m of a body moving with velocity v relative to a stationary observer varies with v and is given by

$$m = \frac{m_0}{\sqrt{\left[1-\left(v^2/c^2\right)\right]}} \quad \text{... (i)}$$

where m_0 is the 'rest mass' of the body and c the velocity of light.

The variation of mass with velocity has modified our ideas about energy. Let us consider a particle of mass m acted upon by a force F in the same direction as its velocity v. The force is defined as the rate of change of momentum *i.e.*

$$F = \frac{d}{dt}(mv) = m\frac{dv}{dt} + v\frac{dm}{dt}. \text{(} m \text{ is variable)}$$

The work done by the force F in a displacement ds of the body is equal to the change in the kinetic energy K of the body. Thus

$$dK = F\ ds = m\frac{dv}{dt}\ ds + v\ \frac{dm}{dt} ds$$

$$= m\ \frac{ds}{dt}\ dv + v\ \frac{ds}{dt}\ dm$$

$$= mv\ dv + v^2\ dm, \qquad \text{... (ii)}$$

because $\quad ds/dt = v.$

Now, let us differentiate eq. (i) :

$$dm = m_0\left(-\frac{1}{2}\right)\left(1-\frac{v^2}{c^2}\right)^{-3/2}\left(-\frac{2v\,dv}{c^2}\right)$$

$$= \frac{m_0}{c^2}\frac{v\,dv}{\left(1-\dfrac{v^2}{c^2}\right)^{3/2}}$$

But $\qquad m_0 = m\left(1-\dfrac{v^2}{c^2}\right)^{1/2}$ by eq. (i).

$$\therefore \qquad dm = \frac{mv\,dv}{\left(c^2-v^2\right)}$$

or $\qquad mv\ dv = (c^2 - v^2)\ dm.$

Making this substitution in eq. (ii), we get

$$dK = (c^2 - v^2)\ dm + v^2\ dm = c^2\ dm.$$

Suppose the body has a mass m_0 when at rest and a mass m when accelerated to a velocity v. The kinetic energy acquired is then

$$K = \int dK = \int_{m_0}^{m} c^2\ dm = c^2\ (m - m_0). \qquad \text{... (iii)}$$

This result shows that the kinetic energy of a body is equal to the relativistic increase in mass of the body over the rest mass multiplied by the square of the velocity of light. It can be interpreted to mean that even when the body is at rest, it possesses an amount of energy w_0c^2. Accordingly, m_0c^2 is called the 'rest energy' E_0 of a body whose rest mass is m_0. The *total* energy E of the body is therefore the sum of the kinetic energy and the rest energy. Thus

$$E = K + E_0 = (m-m_0)\, c^2 + m_0c^2.$$

or

$$E = mc^2$$

This is the famous Einstein's mass-energy relation. This shows that with the mass m of a system is associated an amount of energy mc^2 or, conversely, a system with total energy E has associated with it an inertial mass E/c^2.

Whenever we supply any kind of energy to a body, the mass of the body increases. In fact, a body when heated becomes more massive.

Verification: The mass and energy relation has been verified in a number of phenomena :

(i) *Compton Effect:* Compton treated the X-ray scattering as an elastic collision between a photon and an electron, after which the scattered photon moves in a new direction and the electron recoils with a velocity comparable with the velocity of light. Applying the laws of conservation of energy and momentum in the light of mass-energy relation, Compton calculated a value for the wavelength shift which agreed with the experimental results.

(ii) *Fine Structure of Spectral Lines:* Sommerfeld explained the fine structure of spectral lines on the basis of relativistic variation of mass. The nice agreement of his theory with experiment provides another verification of mass-energy relation.

(iii) *Nuclear Phenomena:* The explanation of 'mass defect' and 'release of tremendous amount of energy in nuclear fission' which is entirely based on mass-energy relation, gives a strong support to the relation.

Compressed Spring Dissolved in Acid : The idea of mass-energy equivalence can be extended to energies other than kinetic. For example, when we compress a spring, we give it elastic potential energy ΔU (*say*) and its mass increases by an amount $\Delta U/c^2$. Thus the mass of a compressed spring is greater than that of an uncompressed one. If the compressed spring is dissolved in acid, the reaction products are of slightly greater mass (though undetectable) than if the spring were uncompressed.

Relativistic Kinetic Energy of a Body

We have deduced in the last question that the relativistic kinetic energy of a particle of rest mass m_0 is given by

$$K = c^2 (m - m_0).$$

But $m = \dfrac{m_0}{\sqrt{\left[1-\left(v^2/c^2\right)\right]}}$, where m is the mass of the particle moving with velocity v. Therefore

$$K = c^2 \left[\frac{m_0}{\sqrt{\left[1-\left(v^2/c^2\right)\right]}} - m_0\right]$$

or
$$K = m_0 c^2 \left[\left(1-\frac{v^2}{c^2}\right)^{-1/2} - 1\right]. \qquad \text{...(i)}$$

This is the relativistic kinetic-energy expression in terms of velocity v.

Expanding $\left(1-\dfrac{v^2}{c^2}\right)^{-1/2}$ by binomial theorem, we get

$$\left(1-\frac{v^2}{c^2}\right)^{-1/2} = 1+\frac{1}{2}\frac{v^2}{c^2}+\frac{3}{8}\frac{v^4}{c^4}+\ldots\ldots$$

For small velocities $v << c$, the terms beyond $\frac{1}{2}\frac{v^2}{c^2}$ are negligible.

Then $$\left(1-\frac{v^2}{c^2}\right)^{-1/2} \approx 1+\frac{1}{2}\frac{v^2}{c^2}.$$

Making this substitution in eq. (i), we obtain

$$K \simeq m_0c^2 \left\{1+\frac{1}{2}\frac{v^2}{c^2}-1\right\} \simeq \frac{1}{2}m_0v^2,$$

which is the classical expression for the kinetic energy. The classical kinetic energy is less than the rest energy (m_0c^2).

We have seen above that the relativistic kinetic energy is

$$K = c^2\left[\frac{m_0}{\sqrt{\left[1-\left(v^2/c^2\right)\right]}}-m_0\right].$$

If $v = c$, then $\sqrt{\left[1-\left(v^2/c^2\right)\right]} = 0$ and so

$$K = \infty.$$

Relativistic Relation between Energy and Momentum: The total (rest-mass plus kinetic) energy of a particle is given by

$$E = mc^2 = \frac{m_0c^2}{\sqrt{\left(1-\frac{v^2}{c^2}\right)}}, \qquad \ldots \text{(i)}$$

and the relativistic momentum is given by

$$p = mv = \frac{m_0v}{\sqrt{\left(1-\frac{v^2}{c^2}\right)}} \qquad \ldots \text{(ii)}$$

and do $$pc = \frac{m_0 vc}{\sqrt{\left(1 - \frac{v^2}{c^2}\right)}}. \qquad ...(iii)$$

Squaring andsubtracting eq. (iii) from eq. (i), we got

$$E^2 - p^2c^2 = \frac{m_0^2c^4 - m_0^2v^2c^2}{1 - \frac{v^2}{c^2}}$$

$$= \frac{m_0^2c^4\left(1 - \frac{v^2}{c^2}\right)}{1 - \frac{v^2}{c^2}} = m_0^2c^4$$

or $$E^2 = m_0^2c^4 + p^2c^2.$$

or $$E = \sqrt{\left(m_0^2\, c^4 + p^2c^2\right)}.$$

This is the relativistic relation between total energy E and momentum p of a particle.

Relativistic Relation between Kinetic Energy and Momentum : The relativistic kinetic energy of a particle is total energy minus rest-mass energy, that is,

$$K = E - m_0c^2 = \sqrt{\left(m_0^2\, c^4 + p^2c^2\right)} - m_0c^2$$

$$= m_0c^2 \left\{\left(1 + \frac{p^2}{m_0^2c^2}\right)^{1/2} - 1\right\}$$

This is the required relation.

When $v < c$, i.e. $p \ll m_o c$, then we can write

$$K \simeq m_o c^2 \left\{\left(1 + \frac{1}{2}\frac{p^2}{m_0^2c^2}\right) - 1\right\}$$

or $$\simeq \frac{p^2}{2m^2}$$

Again, when $v \ll c$, then $m \simeq m_0$.

$$\therefore \quad K \simeq \frac{p^2}{2m}.$$

Thus, in the limit of small velocities, the relativistic between kinetic energy and momentum tends to the classical relation.

Massless Particles

A *massless particle is one which has a zero rest mass.* Classically, such a particle cannot exist. In relativistic mechanics, however, a particle having zero rest mass can exist and show particle-like properties as energy and momentum.

Let us see the requirement for the existence of a massless particle. The relativistic total energy and momentum of a particle of rest mass m_0 moving with speed v are given by

$$E = mc^2 \quad \frac{m_0 c^2}{\sqrt{\left(1 - \frac{v^2}{c^2}\right)}}$$

and $$p = mv = \frac{m_0 v}{\sqrt{\left(1 - \frac{v^2}{c^2}\right)}}.$$

When $m_0 = 0$ and $v < c$, then $E = p = 0$. A massless particle with a speed less than c can have neither energy nor momentum. However, if $m_0 = 0$ and $v = c$, both E and p become indeterminate and can have any values. Hence, *massless particles having energy and momentum can exist provided that they travel with the speed of light.* Photon is an example of massless particle.

Again, the relativistic total energy E of a particle of rest mass in terms of its momentum p is given by

$$E = \sqrt{\left(m_0^2 c^4 + p^2 c^2\right)}$$

If $\qquad m_0 = 0$, then

$$E = pc.$$

This is also the result obtained in classical electromagnetism for radiation (or photon whose rest mass is zero).

Speed of Photon: When a body moves with a speed less than c in one reference frame, we can find another reference frame in which it is at rest. But when it moves with speed c in one reference frame, it will move with the same speed c in all reference frames. Since the photon has a zero rest mass it should always move with the speed of light, whatever is the reference frame.

Transformation of Energy and Momentum

Let there be two reference frames S and S , the frame S' moving with a constant velocity v relative to S. Let u and u' be the velocities of a particle measured by observers in S and S' respectively. By the relativistic addition of velocities, we have

$$u' = \frac{u - v}{1 - \left(uv / c^2\right)} \qquad \text{...(i)}$$

The energy and momentum of the particle in the frame s are

$$E = mc^2 = \frac{m_o c^2}{\sqrt{\left(1 - \frac{u^2}{c^2}\right)}}. \qquad \text{...(ii)}$$

and

$$p = mu = \frac{m_0 u}{\sqrt{\left(1 - \frac{u^2}{c^2}\right)}}. \qquad \text{...(iii)}$$

The corresponding quantities in the frame S′ are

$$E' = mc^2 = \frac{m_0 c^2}{\sqrt{\left(1-\frac{u'^2}{c^2}\right)}}. \qquad \text{...(iv)}$$

and $$p = mu = \frac{m_0 u}{\sqrt{\left(1-\frac{u^2}{c^2}\right)}}. \qquad \text{... (v)}$$

Now, $$\left[1-\frac{u^2}{c^2}\right]^{-1/2} = \left[1-\frac{1}{c^2}\frac{(v-v)^2}{\left(1-\frac{uv}{c^2}\right)^2}\right]^{-1/2}$$

$$= \left[\frac{\left(1-\frac{uv}{c^2}\right)^2 - (u-v)^2/c^2}{\left(1-\frac{uv}{c^2}\right)^2}\right]^{-1/2}$$

$$= \left[\frac{1+\frac{u^2v^2}{c^4}-\frac{2uv}{c^2}-\frac{u^2}{c^2}-\frac{v^2}{c^2}+\frac{2uv}{c^2}}{\left(1-\frac{uv}{c^2}\right)^2}\right]^{-1/2}$$

$$= \left[\frac{\left(1-\frac{u^2}{c^2}\right)-\frac{v^2}{c^2}\left(1-\frac{u^2}{c^2}\right)}{\left(1-\frac{uv}{c^2}\right)^2}\right]^{-1/2}$$

$$= \left[\frac{\left(1-\frac{u^2}{c^2}\right)\left(1-\frac{v^2}{c^2}\right)}{\left(1-\frac{uv}{c^2}\right)^2}\right]^{-1/2}$$

$$= \frac{1 - \dfrac{uv}{c^2}}{\left(1 - \dfrac{u^2}{c^2}\right)^{1/2} \left(1 - \dfrac{v^2}{c^2}\right)^{1/2}} . \qquad \text{...(vi)}$$

We make this substitution in eq. (iv).

$$E' = m_o c^2 \frac{\left(1 - \dfrac{uv}{c^2}\right)}{\left(1 - \dfrac{u^2}{c^2}\right)^{1/2} \left(1 - \dfrac{v^2}{c^2}\right)^{1/2}}$$

$$= \frac{1}{\left(1 - \dfrac{v^2}{c^2}\right)^{1/2}} \left[\frac{m_0 c^2}{\left(1 - \dfrac{u^2}{c^2}\right)^{1/2}} - \frac{m_0 uv}{\left(1 - \dfrac{u^2}{c^2}\right)^{1/2}} \right] .$$

Using eq. (ii) and (iii), we get

$$E' = \frac{E - pv}{\left(1 - \dfrac{v^2}{c^2}\right)^{1/2}} \qquad \text{... (vii)}$$

Again, making substitutions from eq. (vi) and eq. (i) in eq. (v), we get

$$p' = \text{m}_0 \frac{u - v}{1 - \dfrac{uv}{c^2}} \frac{\dfrac{uv}{c^2}}{\left(1 - \dfrac{u^2}{c^2}\right)^{1/2} \left(1 - \dfrac{v^2}{c^2}\right)^{1/2}}$$

$$= \frac{1}{\left(1 - \dfrac{v^2}{c^2}\right)^{1/2}} \left[\frac{m_0 u}{\left(1 - \dfrac{u^2}{c^2}\right)^{1/2}} - \frac{m_0 v}{\left(1 - \dfrac{u^2}{c^2}\right)^{1/2}} \right] .$$

Using eq. (iii) and (ii), we get

$$p' = \frac{p - \frac{Ev}{c^2}}{\left(1 - \frac{v^2}{c^2}\right)^{1/2}}. \quad \text{...(viii)}$$

Eq. (vii) and (viii) are the required expressions,

It is Proved that $E^2 - p^2c^2 = m_0^2\, c^4$ is invariant under Lorentz transformation.

The right-hand side $m_0^2c^4$ is obviously invariant. Hence it is enough to show that

$$E'^2 - p'^2c^2 = E^2 - p^2c^2$$

where E' and p' are the energy and momentum in a frame S' moving uniformly with a velocity v relative to S in which the energy and momentum are E and p.

The Lorentz transformation equations for energy and momentum are

$$E' = \frac{E - pv}{\left(1 - \frac{v^2}{c^2}\right)^{1/2}} \text{ and } p = \frac{p - \frac{Ev}{c^2}}{\left(1 - \frac{v^2}{c^2}\right)^{1/2}}.$$

Now,

$$E'^2 - p'^2\, c^2 = \frac{\left(E^2 + p^2v^2 - 2Epv\right) - \left(p^2 + \frac{E^2v^2}{c^4} - \frac{2pEv}{c^2}\right)c^2}{1 - \frac{v^2}{c^2}}$$

$$= \frac{E^2 + p^2v^2 - 2Epv - p^2c^2 - \frac{E^2V^2}{c^2} + 2pEv}{1 - \frac{v^2}{c^2}}$$

$$= \frac{E^2\left(1-\frac{v^2}{c^2}\right) - p^2c^2\left(1-\frac{v^2}{c^2}\right)}{1-\frac{v^2}{c^2}} = E^2 - p^2c^2.$$

Thus $E^2 - p^2c^2$ is Lorentz invariant.

Doppler Effect in Relativity

Suppose a light source of frequency ν is placed at the origin of S-frame. An observer in S′-frame moving *away* from the source at speed v measures a frequency of ν′. We have to find a relation between ν′ and ν.

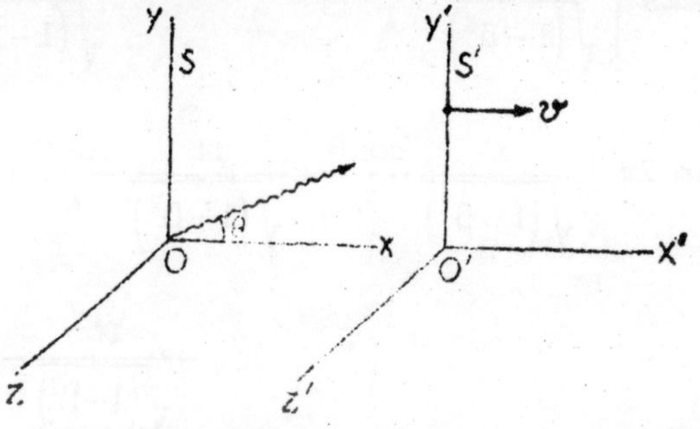

A light wave emitted from the source and moving in the *X-Y* plane making an angle θ with the *x*-axis would be represented

$$\sin\, 2\pi \left(\frac{x\cos\theta' + y\sin\theta'}{\lambda} - v't'\right), \qquad \ldots \text{(i)}$$

where $\lambda v = c$, the velocity of wave propagation. The amplitude of the wave is supposed to be unity.

The expression describing the propagation in the S′- frame would be of the same form :

$$\sin 2\pi \left(\frac{x'\cos\theta + y'\sin\theta'}{\lambda'} - v't'\right), \qquad \ldots \text{(ii)}$$

where λ and are the wavelength and frequency respectively as observed in *S*. Again $\lambda v = c$, which is same for both the frames.

Let us apply Lorentz transformation equations to (i) by putting,

$$x = \frac{x' + vt'}{\sqrt{(1-\beta^2)}}, \; y = y' \text{ and } t = \frac{t' + (x'v/c^2)}{\sqrt{(1-\beta^2)}},$$

where $\beta = v/c$. We obtain

$$\sin 2\pi \left(\frac{(x' + vt')}{\sqrt{(1-\beta^2)}} \frac{\cos\theta}{\lambda} + \frac{y' \sin\theta}{\lambda} - v \frac{t' + (x'v/c^2)}{\sqrt{(1-\beta^2)}} \right)$$

$$= \sin 2\pi \left(\frac{x'}{\sqrt{(1-\beta^2)}} \frac{\cos\theta}{\lambda} + \frac{vt'}{\sqrt{(1-\beta^2)}} \frac{\cos\theta}{\lambda} + \frac{y' \sin\theta}{\lambda} - \frac{vt'}{\sqrt{(1-\beta^2)}} - \frac{(c/\lambda)\, x'v/c^2}{\sqrt{(1-\beta^2)}} \right)$$

$$= \sin 2\pi \left[\frac{x'}{\sqrt{(1-\beta^2)}} \frac{\cos\theta}{\lambda} + \frac{\beta t' v \cos\theta}{\sqrt{(1-\beta^2)}} + \frac{y' \sin\theta}{\lambda} - \frac{vt'}{\sqrt{(1-\beta^2)}} - \frac{\beta x'}{\sqrt{(1-\beta^2)}} \frac{1}{\lambda} \right]$$

$$= \sin 2\pi \left[\frac{x'(\cos\theta - \beta)}{\lambda\sqrt{(1-\beta^2)}} + \frac{y' \sin\theta}{\lambda} - \frac{vt'(1 - \beta\cos\theta)}{\sqrt{(1-\beta^2)}} \right]. \quad \text{(iii)}$$

This must represent a propagation in S-frame, and hence must be identical to eq. (i). Equating the coefficients of t' in (ii) and (iii), we get

$$v' = \frac{v(1 - \beta\cos\theta)}{\sqrt{(1-\beta^2)}}. \qquad ...(iv)$$

This is the general expression for Doppler effect in relativity.

If $\theta = 0$, when the observer is in relative motion along the direction of propagation of light and is moving away from the source, we have

$$v' = \frac{v(1-\beta)}{\sqrt{(1-\beta^2)}} = \frac{v(1-\beta)}{\sqrt{[(1-\beta)(1+\beta)]}}$$

or
$$v' = v\sqrt{\left(\frac{1-\beta}{1+\beta}\right)} = v\sqrt{\left(\frac{c-v}{c+v}\right)}$$

which shows that the observed frequency v′ is lower than the actual frequency v. If the observer is moving towards the source (negative), then we would have

$$v' = v\sqrt{\left(\frac{c+v}{c-v}\right)},$$

which shows that the observed frequency v' is higher than the actual frequency v. This is the "longitudinal" Doppler effect in light.

These results have been confirmed experimentally by Ives and Stilwell in 1938 who used a beam of excited hydrogen atoms of well-defined speed and direction as the source of light.

If the observer is in relative motion perpendicular to the direction of propagation of light, then also a Doppler effect is observed. Let us write eq. (iv) inversely as

$$v = \frac{v'(1 + \beta\cos\theta')}{\sqrt{(1-\beta^2)}}.$$

Now putting θ = 90°, we get

$$v = \frac{v'}{\sqrt{(1-\beta^2)}} \quad \text{or} \quad v' = v\sqrt{(1-\beta^2)}$$

Thus, if our line of sight is 90° to the relative motion, then the observed frequency is lower than the actual frequency v. This is known as 'transverse Doppler effect' and has been confirmed experimentally. This is purely a relativistic phenomenon. There is no transverse Doppler effect in classical Physics ($v<<c$). This can be checked by ignoring β^2 in the last expression which then gives $v' = v$.

General Theory of Relativity: Einstein's special or restricted theory of relativity is based on the assumption that the laws of Physics have the same form in all frames of reference moving with a *constant* velocity with respect to one another (inertial frames). Einstein later on extended his theory to include the accelerated motions also and hence to describe the gravitational phenomena. His first postulate of general relativity states that "the laws of Physics can be expressed in equations having the same form in all frames of reference, regardless of their relative motion."

The second postulate of general relativity states that no experiment performed in a closed laboratory can distinguish between the effects produced by a gravitational field and those produced by an acceleration of the laboratory (relative to stars)." This postulate is known as the "principle of equivalence". For instance, if a man were inside a closed elevator car in outer space (Fig.), and if the car were suddenly accelerated upwards, he can not decide whether a gravitational force had pulled him downwards or the car had been accelerated upwards.

The general theory which have been tested the following predictions

A Ray of Light is bent by the Sun's Gravitational Field: During a solar eclipse a star *S* (Fig.) near the sun appears to be displaced to *S′*. This is because the light from the star passing near the sun is bent in sun's gravitational field.

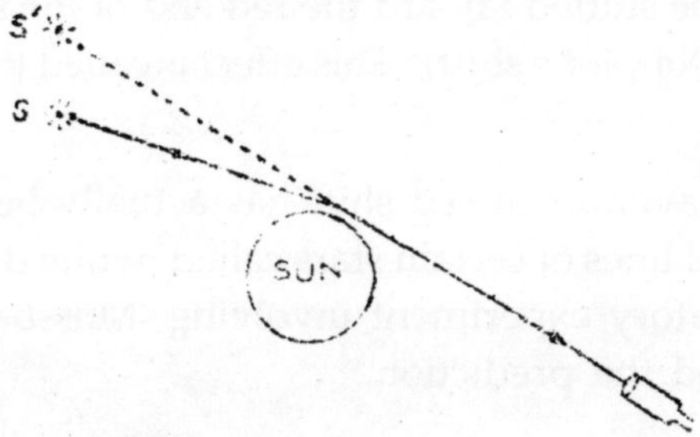

The observed magnitude of this displacement agrees with the value predicted by the general theory within about 10%.

There is an Advance of the Perihelion of Mercury's Orbit: The planet Mercury revolves round the Sun in an elliptic orbit with Sun at one of its foci. The point of the orbit at which the planet is nearest to the Sun is called the 'perihelion'. The perihelion itself is found to be very slowly rotating around the Sun (at a rate of about 2° per century), so that the orbit is as shown in Fig. This effect is called the 'advance of the perihelion' of Mercury. Whereas more than 99% of the observed perihelion advance can be explained as due to the gravitational attraction of the other planets on Mercury; the remaining 'residual' advance can only be explained on the basis of general relativity.

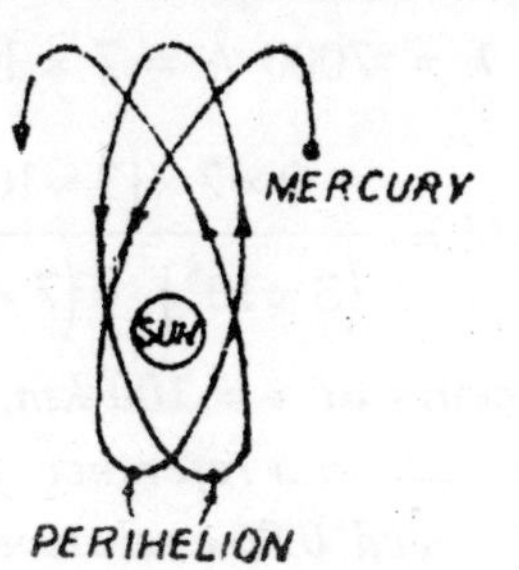

A Stationary Clock runs more Slowly in a Strong Gravitational Field than in a Weak One: An atom-emitting light of a definite frequency is equivalent to a clock (atomic clock). This third prediction means that the light emitted by an atom on a star must be of frequency lower than that emitted by the same atom on earth.* In other words, the light from the star must be shifted toward the red end of the spectrum (apart from any Doppler's shift). This effect is called the 'gravitational red shift'.

The gravitational red shift has actually been observed in the spectral lines of certain stars called 'white dwarfs', and also in a laboratory experiment involving 'Mossbauer effect' and has verified the prediction.

PROBLEMS

1. Let's see the fringe-shift according to the ether theory in the Michelson-Morley experiment, if the effective length of each path is 7 meters and light has 7000 Å wavelength. The velocity of earth is 3 ×10⁴ m/sec.

Solution : The expected shift in the Michelson-Morley experiment is given by

$$\Delta N = \frac{2lv^2}{c^2\lambda}$$

Here $l = 7$ m, $v' = 3 \times 10^4$ m/s,

$c = 3 \times 10^8$ m/s

and $\lambda = 7000$ Å $= 7 \times 10^{-7}$ m.

$$\therefore \quad \Delta N = \frac{2 \times 7 \times \left(3 \times 10^4\right)^2}{\left(3 \times 10^8\right)^2 \times \left(7 \times 10^{-7}\right)} = 0.2.$$

2. An event occurs at x = 100 km, y = 10 km and z = 1.0 km at t = 2.0 × 10⁻⁴ sec in a reference frame S. Another frame S″ is moving with speed 0.95 c relative to S along the common

x-x′ axis, the origins coinciding at t = t′ = 0. We can see the result of computing the coordinates x′, y, z′, t′ of the event in S′.

Solution : By Lorentz transformation, we have

$$x' = \frac{x - vt}{\sqrt{\left[1-(v/c)^2\right]}}$$

$$= \frac{100\text{ km} - \left(0.95 \times 3 \times 10^5\text{ km/sec}\right)\left(2.0 \times 10^{-4}\text{ sec}\right)}{\sqrt{\left[1-(0.95)^2\right]}}$$

$$= \frac{100\text{ km - }57\text{ km}}{\sqrt{(1-0.9025)}} = \frac{43\text{ km}}{\sqrt{(0.0975)}} = \frac{43}{0.312} = 137.8\text{ km}$$

$$y' = y = 10\text{ km.}$$

$$z' = z = 1.0\text{ km.}$$

$$t' = \frac{t-\left(xv/c^2\right)}{\sqrt{\left[1-(v/c)^2\right]}} = \frac{\left(2.0 \times 10^{-4}\text{ sec}\right) - \dfrac{100\text{ km} \times 0.95}{3 \times 10^5\text{ km/sec}}}{\sqrt{\left[1-(0.95)^2\right]}}$$

$$= \frac{2.0 \times 10^{-4}\text{ sec} - 3.166 \times 10^{-4}\text{ sec}}{0.312} = -3.74 \times 10^{-4}\text{ sec.}$$

3. In a reference frame S an event 1 occurs at the origin at t = 0 and another event 2 occurs at x = 3000 meter and t = 4 × lO⁻⁶ sec. Let's find the time-interval bet:veen events as registered by clocks in a frame S moving with a speed v = 0.6 c relative to S along the common x–x′ axis, the origins coinciding t = t′ - 0.

Solution : Let t_1' and t_2' be the times of the events 1 and 2 as registered on S' clocks. Then by Lorentz transformation, we have

$$t_1' = \frac{t_1 - (x_1 v/c^2)}{\sqrt{[1-(v/c)^2]}} = 0$$

and
$$t_2' = \frac{t_1 - x_2 v/c^2}{\sqrt{[1-(v/c)^2]}}$$

$$= \frac{4 \times 10^{-6} \text{ sec} - \dfrac{3000 \text{ m} \times 0.6}{3 \times 10^8 \text{ m/s}}}{\sqrt{[1-(0.6)^2]}}$$

$$= \frac{4 \times 10^{-6} - 6 \times 10^6}{0.8} = -2.5 \times 10^{-6} \text{ sec.}$$

$\therefore$ time-interval $t_2' — t_1' = -2.5 \times 10^{-8}$ sec.

The *S* clock registers the event 2 earlier than *S* clock.

4. *We may find out what would be the apparent length of a meter stick measured by an observer at rest when the stick is moving along its length with a velocity equal to c, c/ √2, √3 c/2, c/2, 0-8 c.*

Solution : As a consequence of Lorentz transformation equations, when a body moves with a velocity v relative to the observer, its measured length is contracted in the direction of its motion by the factor $\sqrt{[1-(v^2/c^2)]}$. Thus if L_0 be the true length (when at rest), the length when in motion is

$$L = L_0 \sqrt{\left(1 - \frac{v^2}{c^2}\right)}.$$

When $v = c$, $v^2/c^2 = 1$, then the length of meter stick (L_o = 100 cm) would be

$$L = L \sqrt{(1-1)} = 0$$

When $$v = \frac{c}{\sqrt{2}}, \frac{v^2}{c^2} = \frac{1}{2}, \text{ then}$$

$$L = L_0 \sqrt{\left(1 - \frac{1}{2}\right)} = \frac{L_0}{\sqrt{2}}$$

$$= \frac{100 \text{ cm}}{1.41} = 70.7 \text{ cm.}$$

When $$v = \frac{\sqrt{3}}{2} c, \frac{v^2}{c^2} = \frac{3}{4}, \text{ then}$$

$$L = L_0 \sqrt{\left(1 - \frac{3}{4}\right)} = \frac{L_0}{2}$$

$$= \frac{100 \text{ cm}}{2} = 50 \text{ cm.}$$

When $$v = \frac{c}{2}, \frac{v^2}{c^2} = \frac{1}{4}, \text{ then}$$

$$L = L_0 \sqrt{\left(1 - \frac{1}{4}\right)} = \frac{\sqrt{3 L_0}}{2}$$

$$= \frac{1.73 \times 100 \text{ cm}}{2} = 86.5 \text{ cm.}$$

When $$v = 0.8\, c, \frac{v^2}{c^2} = 0.64, \text{ then}$$

$$L = L_0 \sqrt{(1-0.64)} = 0.6\, L_0$$

$$= 0.6 \times 100 \text{ cm} = 60 \text{ cm.}$$

5. A spaceship 5.0 meter long passes the earth at a speed of 2.8×10^8 ms^{-1}. Its apparent length will be? (c = 3.0 10^8 ms^{-1}).

Solution : If L_0 be the best length of the spaceship, then its apparel length in light is given by

$$L = L_0 \sqrt{\left[1-\left(v^2/c^2\right)\right]}$$

$$= \text{(50 meter)} \sqrt{[1 - (2.8/3)^2]} = 18 \text{ meter.}$$

6. Let's calculate the length and the orientation of a meter rod in a frame of reference which is moving with a velocity equal to 0.6 c, in a direction making an angle of 30° with the rod.

Solution : The component of the length of the 1- meter rod along the direction of motion of the frame is L_{xo} = 1 cos 30° meter, and that perpendicular to this direction is L_{yo} = 1 sin 30° meter.

The apparent length along the moving frame is

$$L_x = L_{xo} \sqrt{1 - \frac{v^2}{c^2}} = (1 \cos 30°) \sqrt{[-(0.6)^2]}$$

$$= (1 \cos 30°)\ 0.8 = \frac{2}{5} \sqrt{3} \text{ meter.}$$

The length perpendicular to the direction of motion remains the same *i.e.*

$$L_y = L_{yo} = \text{I} \sin 30° = \frac{1}{2} \text{ meter.}$$

∴ length observed in the moving frame

$$L = \sqrt{\left(L_x^2 + L_y^2\right)} = \sqrt{\left(\frac{12}{25} + \frac{1}{4}\right)}$$

$$= \sqrt{(0.73)} = 0.854 \text{ meter.}$$

If the rod appears to make an angle θ with the direction of motion, then

$$\tan\theta = \frac{L_y}{L_x} = \frac{1/2}{\frac{2}{5}\sqrt{3}}$$

$$= \frac{5}{4\sqrt{3}} = 0.72.$$

$$\therefore \qquad \theta = \tan^{-1}(0.72).$$

7. The percentage contraction of a rod moving with a velocity of 0.8 c in a direction inclined at 60° to its own length may be calculated.

Solution : Let L_0 be the length of the rod at rest. Its components along and perpendicular to the direction of motion are $L_0 \cos 60°$ and $L_o \sin 60°$ respectively.

The apparent length along the direction of motion

$$= L_o \cos 60° \sqrt{[1-(0.8)^2]} = 0.3\, L_0$$

and that perpendicular to the direction of motion (no change)

$$= L_0 \sin 60° = \frac{\sqrt{(3)}}{2} L_0$$

$\therefore$ length of the moving rod is

$$L = \sqrt{\left[(0.3\, L_0)^2 + \left(\frac{\sqrt{(3)}}{2} L_0\right)^2\right]} = 0.916\, L_0.$$

Percentage contraction

$$\frac{L_0 - L}{L_0} \times 100 = \frac{L_0 - 0.916\, L_0}{L_0} \times 100 = 8.4\%$$

According to a consequence of Lorentz transformation equations, the length of a body moving with a velocity v relative to the observer is contracted by a factor $\sqrt{(1 - \beta^2)}$ (where $\beta = v/c$) in the direction of its motion, whereas its dimensions perpendicular to the direction of motion are unaffected.

Thus the cube, the rest length of each side of which is L_0, has its length along the direction of motion as

$$L_0 (1- \beta^2)^{1/2},$$

while the lengths in other two directions are still L_o each. Hence the volume when in motion is

$$L_0^3 (1- \beta^2)^{1/2}.$$

9. *A rocket ship is 100 meters long on the ground. When it is in flight its length is 99 meters to an observer on the ground. Its speed may be calculated in following way? (c = 3 × 10⁸ m/sec.)*

Solution : Let L_0 be the length of the rocket ship on the ground. When it is in flight with velocity v, it appears contracted to a length L to an observer on the ground, where

$$L = L_0 \sqrt{[1 - (v^2/c^2)]},$$

where c is the speed of light. From this, we have

$$L - \frac{v^2}{c^2} = \left(\frac{L}{L_0}\right)^2$$

or

$$v^2 = \left[1 - \left(\frac{L}{L_0}\right)^2\right] c^2$$

$$\therefore \quad v = \sqrt{[1 - (L^2/L_0)^2]}\ c.$$

Here $\quad L / L_o = 99/100 = 0.99.$

$$\therefore \quad v = \sqrt{[1 - (0.99)^2]}\ c.$$

Now, $\sqrt{[(1 - 0.99)^2]} = \sqrt{[(1 - 0.99) \times (1 + 0.99)]} = \sqrt{[0.01 \times 1.99]}$

$$= \sqrt{[0.0199]} = 0.141.$$

$$\therefore \quad v = 0.141\ c.$$

The lifetime of a moving particle appears to be lengthened to a stationary observer. If t_0 be the actual lifetime (in the frame of reference of the particle itself), and t that measured by a stationary observer, then

$$t = \frac{t_0}{\sqrt{\left[1 - \left(v^2/c^2\right)\right]}}$$

Here $\quad t_0 = 2.5 \times 10^{-8}$ sec, $v = 2.4 \times 10^{10}$ cm/sec

and we know that $c = 3.0 \times 10^{10}$ cm/sec.

$$\therefore \quad t = \frac{2.5 \times 10^{-8} \text{ sec}}{\sqrt{\left[1 - \left(\frac{2.4 \times 10^{10}}{3.0 \times 10^{10}}\right)^2\right]}} = \frac{2.5 \times 10^{-8} \text{ sec}}{\sqrt{\left[1 - (0.8)^2\right]}}$$

$$= \frac{2.5 \times 10^{-8} \text{ sec}}{0.6} = 4.1 \times 10^{-8} \text{ sec.}$$

Distance travelled by the particle before disintegrating

$= \text{velocity} \times \text{lifetime}$

$= (24 \times 10^{10} \text{ cm/sec}) \times (4.17 \times 10^{-8} \text{ sec})$

$= 1000 \text{ cm.}$

Distance travelled if there were no relativity effect

$= \text{velocity} \times \text{proper lifetime}$

$= (2.4 \times 10^{10} \text{ cm/sec}) \times (2.5 \times 10^{-8} \text{ sec})$

$= 600 \text{ cm.}$

10. Muons have an average life-time of 2 micro-second in their own frame of reference. If a beam of cosmic-ray muons has a speed of 0.98 c relative to earth, find the apparent average life and the distance travelled by the beam during the life-interval; is (C = 3.0 × 10⁸ m/s). 10 micro-second, 2.94 km.

A beam of particles of half-life 2.8 × 10⁻⁸ sec travels in the laboratory with speed 0'96 times the speed of light. The distance the beam travel s before the flux falls to $\frac{1}{2}$ time the initial flux would be (c = 3.0 × 10⁸ m/s):

Solution : The proper half-life of the beam of particles is $t_0 = 2'8 \times 10^{-8}$ sec. The apparent half-life of the beam moving with velocity v as measured by a stationary observer is

$$t = \frac{t_0}{\sqrt{\left[1 - \left(v^2 / c^2\right)\right]}}$$

$$= \frac{2.8 \times 10^{-6}}{\sqrt{\left[1-(0.96)^2\right]}} = \frac{2.8 \times 10^{-6}}{\sqrt{\left[(1-0.96)(1+0.96)\right]}}$$

$$= \frac{2.8 \times 10^{-6}}{\sqrt{(0.04 \times 1.96)}} = \frac{2.8 \times 10^{-6}}{0.28}$$

$$= 1 \times 10^{-5} \text{ sec.}$$

The half-life is defined to be the time during which half of the initial particles decay. Thus, the distance travelled before the flux falls to half

$$= \text{speed} \times \text{half-life (apparent)}$$

$$= 0.96 \text{ c} \times t$$

$$= 0.96 \times (3 \times 10^8 \text{ m/s}) \times (1 \times 10^{-5} \text{ s})$$

$$= 2.88 \times 10^3 \text{ meter} = 2.88 \text{ km.}$$

11. The velocity of π-mesons with proper mean life is 2.5 × 10^−8 sec and observed mean life 2′5 × 10-^7 sec. may be found in following way.

Solution : As a result of relativistic time dilation, the observed mean life is given by

$$t = \frac{t_0}{\sqrt{\left[1-(v^2/c^2)\right]}}$$

where t_0 is the proper life and v the speed of the particle. Thus

$$\sqrt{\left[1-(v^2/c^2)\right]} = \frac{t_0}{t}$$

$$= \frac{2.5 \times 10^{-8}}{2.5 \times 10^{-7}} = 0.1$$

or $$= 1 - \frac{v^2}{c^2} = 0.01$$

or $$\frac{v^2}{c^2} = 1 - 0.01 = 0.99$$

or $$v = \sqrt{(0.99)}\ \text{c} = 0.995\ \text{c}.$$

12. A clock keeps correct time. Let's find out with what speed should it be moved relative to an observer so that it may appear to lose 4 minutes in 24 hours.

If t_0 be the (proper) time-interval on clock at rest relative to an observer and t the time-interval on clock in motion relative to the observer, then

$$t = \frac{t_o}{\sqrt{\left[1-\left(v^2/c^2\right)\right]}},$$

where v is the speed of relative motion.

Here t_0 — 24 × 60 = 1440 min is the proper time-interval measured by an observer moving with the clock and t = 1444 min is the same time-interval measured (with his own clock) by a stationary observer. Thus, to the stationary observer, the moving clock appears to lose 4 min in 24 hours. From the above formula,

$$1444 = \frac{1440}{\sqrt{\left[1-\left(v^2/c^2\right)\right]}}.$$

For $v/c \ll 1, \quad \dfrac{1}{\sqrt{\left(1-\dfrac{v^2}{c^2}\right)}} \approx 1 + \dfrac{v^2}{2c^2}.$

Then

$$1444 = 1440\left(1+\frac{v^2}{2c^2}\right)$$

$$\frac{v^2}{2c^2} = \frac{1444}{1440} - 1 = \frac{4}{1440}$$

$$\text{or} \qquad \frac{v^2}{c^2} = \frac{8}{1440} = 0.0055$$

$$\text{or} \qquad v = \sqrt{(0.0055)}\ c = 0.0745\ c$$

$$= (0.0745)\ (3 \times 18^8\ m/s,$$

$$= 2.23 \times 10^7\ m/s.$$

13. *Two particles are moving in opposite directions each with a speed of 0.9 c in laboratory frame of reference. Let's find the velocity of one particle relative to other.*

Solution : The particles are moving with velocities +0.9 c and –0.9 c in the laboratory frame, say S' frame. Let S be a reference frame in which the particle with velocity – 0.9c is at rest. Then the velocity of S' (laboratory) relative to S is v = 0.9c. Therefore, the particle which in S' has velocity u' = + 9c has a velocity in S given by

$$u = \frac{u' + v}{1 + \frac{u'v}{c^2}} = \frac{+0.9c + 0.9c}{1 + \frac{(0.9c)(0.9c)}{c^2}}$$

$$= \frac{1.8c}{1 + (0.9)^2} = \frac{1.8c}{1.81} = 0.994\ c,$$

which is less than c. If the Galilean velocity transformations were used, we would find that $u = u' + v$ = 0.9c + 0.9c = 1.8 c, which is greater than c and hence impossible.

14. *In the laboratory two particles are observed to travel in opposite directions each with speed 2.8 × 10^{10} cm/sec. After deducing the relative speed of the particles. (c = 3.0 × 10^{10} cm/sec). The answer would be 2.99 × 10^{10} cm/sec.*

Solution : Rockets A and B are moving with velocities 0.Sc and 0.7c in the earth's frame, say S frame. Let 5 be a frame in which A is at rest. Then the velocity of S' (earth) relative to S is v = – 0.8c. Therefore, rocket B which in S' has velocity u' = 0.7c has a velocity in S (observer on A) given by

$$u = \frac{u' + v}{1 + \frac{u'v}{c^2}} = \frac{0.7c + (-0.8c)}{1 + \frac{(0.7c)(-0.8c)}{c^2}}$$

$$= -\frac{0.1\,c}{1 - 0.56} = -0.23\text{ c.}$$

The negative sign indicates that the rocket *B* is approaching *A* with time.

15. A spaceship moving away from the earth with velocity 05 c fires a rocket whose velocity relative to spaceship is 0.8c (a) away from the earth, (b) towards the earth. We can find the velocity of the rocket as observed from the earth in the two cases.

Solution : We may regard earth as the S-frame, the spaceship as the S′-frame, and the rocket as the object whose velocity in the S-frame we seek. Then

$$u' = +0.8\text{ c},\quad v = 0.5\ c$$

$$\therefore \qquad u = \frac{u' + v}{1 + \frac{u'v}{c^2}} = \frac{0.8\,c + 0.5\,c}{1 + (0.8 \times 0.5)} = 0.93\text{ c.}$$

In the second case,

$$u' = -0.8\text{ c.}$$

$$\therefore \qquad u = \frac{u' + v}{1 + \frac{u'v}{c^2}} = \frac{-0.8\,c + 0.5\,c}{1 + \{(-0.8) \times (0.5)\}}$$

$$= -\frac{0.3\,c}{1 - 0.4} = -0.5\text{ c.}$$

16. On the surface of the earth the mass of a man is 100 kg. When he is in a rocket moving with a speed of 4.2 × 10^7^ meters/ sec relative to the earth, let's see what will be his mass as observed by (1) an observer on the earth, (2) an observer in his rocket (c = 3.0 × 10^8^ meters/sec).

Solution : The rest mass of the man is m_0 = 100 kg. The mass as observed by a stationary observer when the man is moving in a rocket is

$$m = \frac{m_0}{\sqrt{\left[1-(v^2/c^2)\right]}} = \frac{100 \text{ kg}}{\sqrt{\left[1-\left(\frac{4.2\times10^7}{3.0\times10^8}\right)^2\right]}}$$

$$= \frac{100 \text{ kg}}{\sqrt{\left[1-(0.14)^2\right]}} = \frac{100 \text{ kg}}{\sqrt{(0.9804)}} = 101 \text{ kg}$$

An observer moving in the rocket will find the mass of the man the same as the rest mass *i.e.* 100 kg, because there is no *relative* motive between the observer and the man.

17. A charged particle shows an acceleration of 4.2 × 10^12 cm/sec^2 under an electric field at low speed. We can compute the acceleration of the particle under the same field when the speed has reached a value 2.88 × 10^10 cm/sec. The speed of light is 3.00 × 10^10 cm/sec.

Solution : The force F on a particle having charge q under an electric field E is

$$F = qE.$$

At low speed the mass of the particle is equal to its rest mass m_0. Therefore, the acceleration at low speed is

$$a_0 = \frac{F}{m_0} = \frac{qE}{m_0} = 4.2 \times 10^{12} \text{ cm / sec}^2$$

When the particle attains a speed 2.88×10^{10} cm/sec which is comparable to the speed of light, its mass increases to m, where

$$m = \frac{m_0}{\sqrt{\left[1-(v^2/c^2)\right]}}$$

$$= \frac{m_0}{\sqrt{\left[1 - \left(\frac{2.88 \times 10^{10}}{3.00 \times 10^{10}}\right)^2\right]}} = \frac{m_0}{0.28}.$$

Now the acceleration would be

$$a = \frac{F}{m} = 0.28 \frac{F}{m_0}$$

$$= 0.28 \times (4.2 \times 10^{12}) = 1.176 \times 10^{12} \text{ cm/sec}^2.$$

18. The rest mass of an electron is 9 × 10^{-31} kg. We may find out what will be its mass if it were moving with 4/5th the speed of light

Solution : The mass of a body (rest mass m_0) moving with velocity v is given by

$$m = \frac{m_0}{\sqrt{\left[1 - (v^2/c^2)\right]}}$$

Here $\frac{v}{c} = \frac{4}{5}$, and so $\sqrt{[1 - (v^2/c^2)]}$

$$= \sqrt{[1 - (16/25)]} = 3/5.$$

$$\therefore \quad m = \frac{9 \times 10^{-31} \text{ kg}}{3/5} = 15 \times 10^{-31} \text{ kg}.$$

19. The mass and momentum of a proton of rest mass 1.67 × 10^{-24} gm, moving with a velocity of 08c may be calculated. If it collides and sticks to a stationary nucleus of mass 5.0 × 10^{-23} gm, we may find the velocity of the resultant particle ($c = 3 \times 10^{10}$ cm/sec).

[**Ans.** 2.78×10^{-24} gm, 6.67×10^{-14} gm-cm/sec, 1.26×10^9cm/sec].

20. The velocity at which the mass of a particle becomes 1.25 times its rest mass ($c = 3 \times 10^8$ m/s) may be deduced.

Solution: $m = \dfrac{m_0}{\sqrt{\left[1-\left(v^2/c^2\right)\right]}}$.

For $\dfrac{m}{m_0} = 1.25$, we have

$$\frac{1}{\sqrt{\left[1-\left(v^2/c^2\right)\right]}} = 1.25 = \frac{5}{4}$$

or $$1 - \frac{v^2}{c^2} = \frac{16}{25}$$

or $$\frac{v^2}{c^2} = 1 - \frac{16}{25} = \frac{9}{25}.$$

This gives v =0.6c = 1.8×10^8 meter/sec.

21. The result of deducing the velocity at which the mass of a body becomes 4 times of the rest mass will be 2.90×10^8 meters/sec.

Deduce the rest energy of an electron in joules and in electron-volts ($m = 9.1 \times 10^{-31}$ kg, $c = 3.0 \times 10^8$ m/sec). Also deduce the speed at which the total relativistic energy becomes 1.25 times the rest energy.

Solution : The rest energy is

$$E_o = m_o c^8$$

$$= (9.1 \times 10^{-31} \text{ kg}) \times (3.0 \times 10^8 \text{ m/sec})^2$$

$$= 8.19 \times 10^{-14} \text{ joule.}$$

Now 1 electron-volt

$$= 1.6 \times 10^{-19} \text{ joule.}$$

$$\therefore \qquad E_0 = \frac{8.19 \times 10^{-14}}{1.6 \times 10^{-19}}\ 0.51 \times 10^6 \text{ eV}$$

$$= 0.51 \text{ MeV}.$$

The ratio of the total relativistic energy E to the rest energy E_0 is given as 1.25. That is,

$$\frac{E}{E_0} = \frac{mc^2}{m_0c^2} = \frac{m}{m_0} = 1.25.$$

But
$$\frac{m}{m_0} = \frac{1}{\sqrt{\left[1-\left(v^2/c^2\right)\right]}}.$$

$$\therefore \qquad \frac{1}{1-\left(v^2/c^2\right)} = (1.25)2 = 1.5625$$

or
$$1 - \frac{v^2}{c^2} = \frac{1}{1.5625}$$

or
$$\frac{v^2}{c^2} = 1 - \frac{1}{1.5625} = \frac{0.5625}{1.5625}$$

$$\therefore \qquad v = \sqrt{\left(\frac{0.5625}{1.5625}\right)}\ c = 0.6 \text{ c}.$$

21. Let's show that the mass of an electron is equivalent to 0.51 MeV energy. The minimum energy of y-ray photon which can produce an electron-positron pair is given below

The rest mass m_0 of electron is 9 I x 10^{-3}) kg. As shown in the last problem, its energy equivalent, is

$$E_0 = 0.51 \text{ MeV}.$$

To produce an electron-positron pair, the minimum energy of the *y*-ray photon is the sum of the test-mass energies of an electron and a positron, that is,

$$0.51 + 0.51 = 1.02 \text{ MeV}.$$

22. *A particle of rest mass m_0 moves with speed $c/\sqrt{2}$. Its mass, momentum, total energy and kinetic energy may be calculated.*

Solution : The relativistic mass of" the particle is

$$m = \frac{m_0}{\sqrt{\left[1-\left(v^2/c^2\right)\right]}} = \frac{m_0}{\sqrt{\left(1-\frac{1}{2}\right)}} = \sqrt{2}\, m_0$$

$$= 1.41\ m_0.$$

The momentum is given by

$$p = mv = \sqrt{2m_0} \times \frac{c}{\sqrt{2}} = m_0\ c.$$

The total energy is

$$E = me^2 = 1.41\ \mathrm{m_0c^2}.$$

The kinetic energy is

$$K = E - m_0c^2 = 0.41\ \mathrm{m_0c^2}.$$

23. *The kinetic energy of a proton (rest mass 1.67 × 10^{-27} kg) moving with velocity 27 × 10^8 m/s is given below. Express the result in MeV. (c = 3.0 × $l0^8$ m/s).*

Solution : The velocity of the proton is comparable to the velocity of light. The (relativistic) kinetic energy is

$$\mathrm{K} = m_0c^2 \left\{\frac{1}{\sqrt{\left[1-\left(v^2/c^2\right)\right]}} - 1\right\}.$$

$$\text{Here} \qquad v/c = \frac{2.7\times10^8\ \mathrm{m/s}}{3.0\times10^8\ \mathrm{m/s}} = 0.9. \text{ Thus}$$

$$\sqrt{[1-(v^2/c^2)]} = \sqrt{[1-(0.9)^2]} = \sqrt{[0.19]} = 0.436.$$

$$\therefore \quad K = (1.67 \times 10^{-27} \text{ kg})$$

$$(3.0 \times 10^8 \text{ m/s})^2 \left\{\frac{1}{0.436} - 1\right\}$$

$$= 1.94 \times 10^{-10} \text{ joule.}$$

But $\quad 1 \text{ MeV} = 1.60 \times 10^{-13}$ joule

$$\therefore \quad K = \frac{1.94 \times 10^{-10}}{1.60 \times 10^{-13}} = 1212 \text{ MeV.}$$

24. The volume of work must be done in order to increase the speed of an electron from 1.8 × 10⁸ m/s to 2.4 × 10⁸ m/s is as follows.

Solution : The initial (relativistic) kinetic energy of the electron (rest mass $m_0 = 9.1 \times 10^{-31}$ kg) is

$$K_1 = m_0c^2 \left\{\left(1 - \frac{v^2}{c^2}\right)^{-1/2} - 1\right\}$$

$$= m_0c^2 \left[\left\{1 - \frac{(1.8 \times 10^8)^2}{(3 \times 10^8)}\right\}^{1/2} - 1\right]$$

$$= m_0c^2 \; [(0.64)^{-1/2} - 1] = 0.25 \; m_0c^2.$$

The final energy is

$$K_2 = m_0c^2 \left[\left\{1 - \frac{(2.4 \times 10^8)^2}{(3 \times 10^8)^2}\right\}^{-1/2} - 1\right]$$

$$= m_oc^2 \; [(0.36)^{-1/2} - 1]$$

$$= 0.67 \; m_0c^2.$$

The work required is equal to the increase in kinetic energy.

Thus

$$W = K_2 - K_1$$

$$= 0.67\ m_0c^2 - 0.25\ m_0c^2 = 0.42\ m_0c^2$$

$$= 0.42 \times (9.1 \times 10^{-31}) \times (3 \times 10^8)^2$$

$$= 34.4 \times 10^{-15} \text{ joule}$$

$$= \frac{34.4 \times 10^{-15}}{1.6 \times 10^{-13}} = 0.215 \text{ MeV.}$$

24. Following is the procedure and result of deducing the speed of an electron accelerated through a potential difference of 10 million volt.

Solution : According to classical mechanics, the kinetic energy of a particle of mass m_0 moving with speed v is mass m_0v^2. When an electron accelerated through a p. d. of V volt acquires this speed, we have

$$\frac{1}{2} m_o v^2 = eV,$$

where e is the electron charge. From this, we get

$$v = \sqrt{\left(\frac{2eV}{m_0}\right)}.$$

Putting $e = 1.6 \times 10^{-19}$ coul, $V =. 10^8$ volts

and $m_0 = 9.1 \times 10^{-31}$ kg, we get

$$v = \sqrt{\left(\frac{2 \times 1.6 \times 10^{-19} \times 10^6}{9.1 \times 10^{-31}}\right)}$$

$$= 5.9 \times 10^8 \text{ meters/sec.}$$

This speed is *greater* than the speed of light (3.0×10^8 m/s) and hence unacceptable, because no particle can have a speed greater than the speed of light. Hence, in the given problem, we must use *relativistic* expression of kinetic energy to calculate the speed of the electron. The expression is

$$K = mc^2 - m_0c^2 = \frac{m_0c^2}{\sqrt{\left[1-\left(v^2/c^2\right)\right]}} - m_0c^2$$

$$= m_0c^2\left\{\frac{1}{\sqrt{\left[1-\left(v^2/c^2\right)\right]}} - 1\right\}.$$

When an electron acquires this energy by means of a potential V, we have

$$m_0c^2\left\{\frac{1}{\sqrt{\left[1-\left(v^2/c^2\right)\right]}} - 1\right\} = \text{eV}$$

or $$\frac{1}{\sqrt{\left[1-v^2/c^2\right]}} = 1 + \frac{eV}{m_0c^2}$$

or $$\frac{1}{\sqrt{\left[1-v^2/c^2\right]}} = 1 + \frac{1.6 \times 10^{-19} \times 10^6}{\left(9.1 \times 10^{-31}\right)\left(3 \times 10^8\right)^2} = 2.95$$

or $$\sqrt{\left[1-\left(v^2/c^2\right)\right]} = \frac{1}{2.96} = 0.339$$

Solving : $v = 0.94\ c = 2.82 \times 10^8$ m/s

25. The speed of an electron which has kinetic energy 102 MeV may be calculated. The rest mass of the electron is 0.51 MeV. (c = 3 × 10^10 cm/sec).

The relativistic expression for the kinetic energy of a particle is $m_0c^2\left\{\frac{1}{\sqrt{\left[1-v^2/c^2\right]}} - 1\right\}$ When an electron acquires it under a potential of V volts, we have

$$m_0c^2 \left\{ \frac{1}{\sqrt{\left[1 - v^2/c^2\right]}} - 1 \right\} = eV,$$

so that $$\frac{1}{\sqrt{\left[1 - v^2/c^2\right]}} = 1 + \frac{eV}{m_0c^2}.$$

The rest mass of the electron is given in energy units (electron-volts); this means that we are given the rest mass *energy* (m_0c^2) of the election. Thus here

$$eV = 1.02 \text{ MeV}$$

and $$m_0c^2 = 0.51 \text{ MeV}.$$

$$\therefore \quad \frac{1}{\sqrt{\left[1 - \left(v^2/c^2\right)\right]}} = 1 + \frac{1.02\ MeV}{0.51\ MeV}$$

$$= 1 + 2 = 3.$$

or $$1 - \frac{v^2}{c^2} = \left(\frac{1}{3}\right)^2 = \frac{1}{9}$$

or $$\frac{v^2}{c^2} = 1 - \frac{1}{9} = \frac{8}{9}$$

or $$\frac{v}{c} = 0.943.$$

$$\therefore \quad v = 0.943 \text{ c} = 0.943 \times (3 \times 10^{10})$$

$$= 2.83 \times 10^{10} \text{ cm/sec.}$$

26. Total energy of an electron (rest mass 9-1 × 10^{-31} kg) moving with speed 0.99 c is given below. Let's find the ratio of Newtonian kinetic energy to the relativistic energy, (c = 3.0 × 10^8 m/see)

Solution : The mass of the electron (rest mass m_0 = 9.1 × 10^{-31} kg) moving with velocity v (= 0.99 c) is

$$m = \frac{m_0}{\sqrt{\left[1-\left(v^2/c^2\right)\right]}} = \frac{9.1 \times 10^{-31}}{\sqrt{[1-(0.99)^2]}}$$

$$= \frac{9.1 \times 10^{-31}}{\sqrt{(1-0.98)}} = \frac{9.1 \times 10^{-31}}{0.141}$$

$$= 64.5 \times 10^{-31} \text{ kg.}$$

The total energy of the electron is therefore

$$E = mc^2 = (64.5 \times 10^{-31} \text{ kg}) (3.0 \times 10^8 \text{ m/sec})^2$$

$$= 5.8 \times 10^{-13} \text{ joule.}$$

The Newtonian kinetic energy is $\frac{1}{2} m_0 v^2$, while the relativistic kinetic energy is total energy *minus* rest energy. Thus their ratio is

$$\frac{\frac{1}{2} m_0 v^2}{mc^2 - m_0 c^2} = \frac{1}{2} \frac{m_0}{m - m_0} \left(\frac{v}{c}\right)^2$$

$$= \frac{1}{2} \frac{9.1 \times 10^{-31}}{55.4 \times 10^{31}} (0.99)^2 = 0.08.$$

27. (i) Let's see what p.d. will accelerate electrons to the speed of light, according to classical Physics (ii), with this p.d., what speed would an electron acquire relativistically (iii) and as to what would be the mass and kinetic energy of the electrons at this speed . The rest mass of the electron is 9.1 × 10^{-31} kg and speed of light is 3 × 10^8 meters/sec. Electronic charge is 1.6 × 10^{-19} coul.

Solution: (i) Classically, the kinetic energy of electron of (rest) mass m_0 at speed c would be $\frac{1}{2} m_0 c^2$. If V is the voltage required to accelerate the electron to speed c, then we have

$$\frac{1}{2} m_0 c^2 = eV. \qquad ...(i)$$

$$\therefore \qquad V = \frac{m_0 c^2}{2e}$$

$$= \frac{(9.1 \times 10^{-31} \text{ kg})(3 \times 10^8 \text{ m/sec})^2}{2(1.6 \times 10^{-19} \text{ coul})}$$

$$= 2.56 \times 10^5 \text{ volts.}$$

(ii) Let v be the speed acquired by the electron relativistically by the p.d. V. Equating relativistic expression for kinetic energy to eV, we have

$$m_0 c^2 \left(\frac{1}{\sqrt{[1-(v/c)^2]}} - 1 \right) = eV$$

Here $eV = \frac{1}{2} m_0 c^2$ by eq (i). Therefore

$$m_0 c^2 \left(\frac{1}{\sqrt{[1-(v/c)^2]}} - 1 \right) = \frac{1}{2} m_0 c^2$$

$$\text{or} \qquad \frac{1}{\sqrt{[1-(v/c)^2]}} = \frac{1}{2} + 1 = \frac{3}{2} \text{ or } 1 - \frac{v^2}{c^2} = \frac{4}{9}$$

$$\text{or} \qquad \frac{v^2}{c^2} = 1 - \frac{4}{9} = \frac{5}{9} = 0.555$$

$$\therefore \qquad v = \sqrt{(0.555)}\, c = 0.745\, c.$$

(iii) The mass of the electron at speed v is given by

$$m = \frac{m_0}{\sqrt{[1-(v/c)^2]}} = \frac{3}{2} m_0$$

$$= \frac{3}{2} \times 9.1 \times 10^{-31} = 1.365 \times 10^{-30} \text{ kg.}$$

The kinetic energy is

$$K = (m - m_0)\ c^2 = \frac{1}{2} m_0c^2$$

$$= \frac{1}{2} \times 9.1 \times 10^{-31} \times (3 \times 10^8)^2$$

$$= 4.09 \times 10^{-14} \text{ joule.}$$

28. The total energy of a particle is exactly twice its rest energy. Its speed is given below

Solution : The total energy E_0 of a moving particle is mc^2 while its rest energy E_0 is m_0c^2. When $E = 2E_0$, then

$$mc^2 = 2m_0c^2 \text{ or } m = 2m_0.$$

The mass of the moving particle is related to its rest mass m_0 by

$$m = \frac{m_0}{\sqrt{\left[1 - \left(v^2 / c^2\right)\right]}}$$

Here $m = 2m_0$.

$$\therefore \qquad 2m_0 = \frac{m_0}{\sqrt{\left[1 - \left(v^2 / c^2\right)\right]}}$$

or
$$1 - \frac{v^2}{c^2} = \frac{1}{4}$$

or
$$v = \frac{\sqrt{3}}{2} c = 0.866 \text{ c.}$$

This result does *not* depend upon the rest mass of the particle.

29. Let's see the speed of an electron whose kinetic energy is equal to its rest energy we may see whether it depends upon the mass of the electron.

Let's calculate the speed of a particle of rest mass 3.'33 × 10⁻²⁷ gm whose energy is 2 MeV.

Solution : The (total) energy of a particle of rest mass m_0 and moving with a speed v is given by

$$E = mc^2 = \frac{m_0c^2}{\sqrt{\left[1-\left(v^2/c^2\right)\right]}}$$

Here $E = 2 \text{ MeV} = 2 \times 1.6 \times 10^{-13}$ joule

and $m_0 = 3.33 \times 10^{-30}$ kg.

$$\therefore \quad 2 \times 1.6 \times 10^{-13} = \frac{\left(3.33 \times 10^{-30}\right)\left(3.0 \times 10^{8}\right)^2}{\sqrt{\left[1-\left(v^2.c^2\right)\right]}}$$

or $$\sqrt{\left[1-\left(v^2/c^2\right)\right]} = \frac{30 \times 10^{-14}}{2 \times 1.6 \times 10^{-13}} = 0.937$$

or $$1-\frac{v^2}{c^2} = (0.937)^2 = 0.878$$

or $$\frac{v^2}{c^2} = 1- 0.878 = 0.122$$

or $$\frac{v}{c} = \sqrt{(0.122)} = 0.349$$

or $$v = 0349 \text{ c} = 0.349 \times (3 \times 10^8 \text{ m/s})$$

$$= 1.047 \times 10^8 \text{ m/s.}$$

30. *A body of mass m_0 at rest breaks up spontaneously into two parts, having rest masses m_{01} and m_{02} and speeds v_1 and v_2 respectively. Show that $m_0 > m_{01} + m_{02}$*

Solution : The total energy of the body at rest is its rest mass energy, m_0c^2. It breaks up into two moving parts. The energies (kinetic + rest mass) of the parts are $\dfrac{m_{01}c^2}{\sqrt{\left[1-\left(v_1/c\right)^2\right]}}$

and $\dfrac{m_{02}c^2}{\sqrt{\left[1-\left(v_2/c\right)^2\right]}}$. By conservation of mass-energy, we have

$$m_o c^2 = \frac{m_{01}\,c^2}{\sqrt{\left[1-(v_1/c)^2\right]}} + \frac{m_{02}\,c^2}{\sqrt{\left[1-(v_2/c)^2\right]}}.$$

Both $\sqrt{[1-(v_1/c)^2]}$ and $\sqrt{[1-(v_1/c)^2]}$ are less than 1.

$\therefore\ m_0 > m_{01} + m_{02}$.

31. ***Two particles, each having rest mass m_0, approach each other with speeds 0.8 c and 0.6 c. They collide and stick together. Compute momentum, total energy, total rest mass, kinetic energy after collision.***

$$\left[\text{Ans. } \frac{7}{12} m_0 c\,,\ \frac{35}{12} m_0 c^2,\ \frac{34.29}{12} m_0,\ \frac{0.71}{12} m_0 c.\right]$$

32. ***A 1-kg mass of iron is heated through 1000°C. Find the increase in its rest mass. Specific heat of iron is 500 joule per kg per °C and the velocity of light is 3 × 10^8 meter/sec.***

Solution : The heat energy absorbed by the iron is

$$\begin{aligned}\Delta E &= \text{mass} \times \text{sp. heat} \times \text{temp-rise} \\ &= 1 \text{ kg} \times 500 \text{ joule/ (kg– °C)} \times 1000°\text{C} \\ &= 5 \times 10^5 \text{ joules.}\end{aligned}$$

$\therefore$ increase in mass,

$$\Delta m = \frac{\Delta E}{c^2} = \frac{5 \times 10^5}{\left(3 \times 10^8\right)^2} = 5.6 \times 10^{-12} \text{ kg}$$

33. ***How much mass is lost when 1 kg of water at 0°C turns to ice at 0°C ? (Latent heat of ice is 89 cal/gm and 1 cal = 4.18 × 10^7 ergs; c = 3 X 10^{10} cm/sec).***

Solution : The heat (energy) withdrawn from the water to freeze it at 0°C is

$$\begin{aligned}\Delta E &= mL \\ &= 1000\text{gm} \times 80\text{cal / gm} = 8 \times 10^4 \text{ cal} \\ &= (8 \times 10^4) \times (4.18 \times 10^7) = 3.344 \times 10^{12} \text{ ergs.}\end{aligned}$$

Let Δm be the corresponding loss in mass. From Einstein's mass-energy equivalence, we have

$$\Delta E = \Delta m\, c^2.$$

$$\therefore \quad \Delta E = \frac{\Delta E}{c^2} = \frac{3.344 \times 10^{12}}{\left(3 \times 10^{10}\right)^2} = 3.72 \times 10^{-9} \text{ gm}$$

34. It would be interesting to see as to how much energy will be obtained if 3.6 gm of mass be completely converted into energy.

Solution : By mass-energy relation, we have

$$\Delta E = (\Delta m)\, c^2.$$

The mass annihilated, Δm = 3.6 gm = 3.6×10^{-3} kg and speed of light $c = 3 \times 10^8$ meter/sec.

$$\therefore \quad \Delta E = (3.6 \times 10^{3})\,(3.0 \times 10^{8})^2$$

$$= 32.4 \times 10^{13} \text{ joule.}$$

Now 1 MeV = 160×10^{-13} joule.

$$\therefore \quad \Delta E = \frac{32.4 \times 10^{13}}{1.6 \times 10^{-13}} = 2.0 \times 10^{27} \text{ MeV.}$$

35. If 4.4 gm of mass be completely converted into energy, the amount of kilowatt-hour electricity could be obtained from it is given below.

Solution : The energy obtained is given by

$$\Delta E = (\Delta m)\, c^2$$

$$= (4.4 \times 10^{-3} \text{ kg})\,(3.0 \times 10^8 \text{ m/s},^2$$

$$= 3.96 \times 10^{14} \text{ joule.}$$

Now 1 kilowatt hour = (1000 × 3600) watt-sec = 3.6×10^6 joule.

$$\therefore \quad \Delta E = \frac{3.96 \times 10^{14}}{3.6 \times 10^{6}} = 1.1 \times 10^8 \text{ kilowatt-hour.}$$

36. The mass equivalent of the energy from an antenna radiating 10000 watts for 24 hours is as follows.

Solution : The total energy radiated is

$$\Delta E = 10000 \times 24 \text{ watt-hours}$$

$$= 10000 \times 24 \times 3600 \text{ watt-sec}$$

$$= 8.64 \times 10^8 \text{ joule.}$$

By Einstein's relation, the mass equivalent is

$$\Delta m = \Delta E/c^2$$

$$= \frac{8.64 \times 10^8 \text{ joule}}{(3.0 \times 10^8 \text{ m/s})^2} = 9.6 \times 10^{-8} \text{ kg.}$$

38. The mass of deuteron is less than the sum of the masses of one proton and one neutron by 0.0024 amu where 1 amu = 1.67 × 10⁻²⁷ kg. Deduce the energy given off (as y-radiation) when a neutron and a proton combine to form a deuteron. We may check in which way the case of fission of U²³⁵ nucleus differs from this case ?

Solution : The mass disappeared during the formation of deuteron is

$$\Delta m = 0.0024 \text{ amu} = 0.0024 \times (167 \times 10^{-27}) \text{ kg}$$

$$= 00040 \times 10^{-27} \text{ kg.}$$

The corresponding energy given off is

$$\Delta E = \Delta m \ c^2$$

$$= (0.0040 \times 10^{-27} \text{ kg}) \times (3 \times 10^8 \text{ meters/sec})^2$$

$$= 3.6 \times 10^{-13} \text{ joule.}$$

In this case the released energy goes as *y*-rays. In nuclear fission a heavy nucleus (such as U^{239}) is split into lighter nuclei. In this process the binding energy per nucleon increases and results in a general release of huge amount of energy. Most of the energy released goes into the kinetic energy of the fission fragments, while about 20% goes as neutrons, γ-rays, neutrinos, etc.

39. It is assumed that sun gets energy by fusion of four hydrogen atoms into a Belgium atom. The rest mass of hydrogen and helium atoms are 10081 and 40039 amu respectively. Let's calculate the energy released in each fusion process. (1 amu -= 1.67 × 10^{-27} kg).

Solution : Mass *of* 4 hydrogen atoms = 4 × 1.0081

$$= 4.0324 \text{ amu.}$$

Mass of I helium atom = 4 0039 amu.

∴ mass-loss in each fusion process

$$\Delta m = 4.0324–4.0039 = 0.0285 \text{ amu}$$
$$= 0.0\ 285 \times (1.67 \times 10^{-27})$$
$$= 4.76 \times 10^{-29} \text{ kg.}$$

∴ energy released

$$\Delta E = \Delta m\ c^2$$
$$= (4.76 \times 10^{-29} \text{ kg}) - (3.00 \times 10^{8} \text{ meters/sec})^2$$
$$= 4.28 \times 10^{-12} \text{ joule.}$$

Now 1 MeV = 1.6 × 10^{-13} joule

$$\therefore \quad \Delta E = \frac{4.28 \times 10^{-12}}{1.6 \times 10^{-13}} = 26.7 \text{ MeV.}$$

40. Masses of electron; H^1 atom and He^4 atom are 000055, 1 00814 and 400387 amu, where 1 amu = l.67 × 10^{-27} kg. Compute the energy (in joules) released when 1 kg hydrogen fuses into helium.

[**Hint :** 1 kg of hydrogen has 6.02 × 10^{26} atoms].

The Answer is 6.46 × 10^{14} Joules.

Bibliography

Anotola, S.A. : *Advanced Physics,* Longman, London, 1968.

Atherton, J.C. : *Thermal Physics,* Butterworths, London, 1987.

Atiyah, M. : *The Geometry and Physics of Knots,* Cambridge, New York, 1991.

Bemier, G. and Kinet J.M. : *Thermal and Statistical Physics,* C.R.C. Press, Boca Raton, 1981.

Benjamin, W.A. : *Problems in Physics,* Academic Press, New York, 1970.

Bowling, D.J.F. : *Basic Physics,* Chapmann and Hall, London, 1976.

Caro, D. E. and Others : *Modern Physics : An Introduction to Atomic and Nuclear Physics,* Edward Arnold, London, 1985.

Cartwright, H.R. : *Elements of Thermal and Statistical Physics,* Oxford University Press, London, 1966.

Clinton, Case M. : *Physical Principles of Flow in Unsaturated Porous Media,* Oxford, New York, 1994.

Crafts, A.S. : *Phloem Transport in Plants,* W.H. Freeman, San Francisco, 1973.

Douglas, J.S. : *Advanced Guide to Modern Physics,* Pelham Books, London, 1976.

Epstein, E. : *Statistical Physics,* John Wiley and Sons, New York, 1972.

Esau, K. : *Physical Geology,* John Wiley and Sons, New York, 1965.

Evans, L.T. : *Thermal Analysis of Materials,* W.A. Benjamin, California, 1975.

Feynman and Others : *The Feymman Lectures on Physics,* Add. Wesley, London, 1989.

Feynman, R. P. and Others : *Elementary Particles and the Laws of Physics,* Cambridge, New York, 2002.

Ftzler M.E. : *Plant Lectins : Molecular and Biological Aspects,* Academic Press, New York, 1985.

Fujimoto, Minoru : *The Physics of Structural Phase Transitions,* Springer, London, 2000.

Galliard, T. and Mercer, E.I. : *Recent Advances in the Chemistry and Biochemistry of Plant Lipids,* Academic Press, New York, 1975.

Green, J.R. : *Modern Physics,* MacMillan, London, 1911.

Haasen, Peter : *Physical Metallurgy,* Cambridge, New York, 1996.

Halevy, A.H. : *Handbook of Physics,* C.R.C. Press, Boca Raton, 1985.

Hopkins, C.R. : *Structure and Function of Cells,* W.B. Saunders Co., London, 1978.

Hornberger, George M. : *Elements of Physical Hydrology,* Johns Hopkins, New York, 1998.

Jorgensen, T. P. : *The Physics of Golf,* Springer, New York, 1999.

Kirkby, E.A. : *Mathematical Physics,* The University of Leeds, Leeds, 1970.

Kuhn, Karl F. : *Basic Physics,* John Wiley, London, 1996.

Kumar, Arun : *Problems in Physics,* Atlantic Publishers, New Delhi, 2003.

Landshoff, P. and Others : *Essential Quantum Physics,* Cambridge, New York, 2001.

Leo, William R. : *Techniques for Nuclear and Particle Physics Experiments : A How to Approach*, Springer, London, 1994.

Lipson, S. G. and Others : *Optical Physics*, Cambridge, New York, 1995.

Lockett, Keith : *Physics in the Real World*, Cambridge, New York, 1995.

Lui, Lam : *Introduction to Nonlinear Physics*, Springer, New York, 1997.

Main, Lain G. : *Vibrations and Waves in Physics*, Cambridge, New York, 1998.

McKee, H.S. : *University Physics*, Oxford University Press, London, 1962.

Meidner, H. and Sheriff, D.W. : *Water and Plants*, Blackie and Sons, Glasgow, 1976.

Mengel, K. and Kirkby, E.A. : *Thermal and Statistical Physics*, Praeger Publishers, New York, 1974.

Nabapro, F. R. N. and Others : *The Physics of Creep*, T & F, New York, 1995.

Negele, J. W. and Others : *Advances in Nuclear Physics*, Plenum, London, 1994.

Palladin, V.I. : *Physical Metallurgy*, Arihant Publishers, Jaipur, 1988.

Plischke, M. and Others : *Equilibrium Statistical Physics*, World Scientific, New York, 1994.

Rabinowitch, E. : *Thermodynamics and Statistical Physics*, John Wiley, New York, 1969.

Rawn, J.D. : *Encyclopaedia of Modern Physics*, Harper and Row, New York, 1983.

Roberts, J.A. and Tucker, G.A. : *Physics and Chance*, Oxford, New York, 1999.

Rossi, Paolo : *The Birth of Modern Science*, Blackwell, New Jersey, 2001.

Ruhland, W. : *Encyclopaedia of Thermodynamics Physics*, Springer-Verlag, Berlin, 1961.

Sabnis, D.D. and Hart, J.W. : *Encyclopaedia of Stastical Physics*, Springer Verlag, Berlin, 1982.

Sagdeev, R. Z. : *Nonlinear Space Plasma Physics*, AIP, New Jersey, 1993.

Salisbury, F.B. and Ross, C.W. : *Thermal Physics*, Woodworth Publishing Co., California, 1967.

Schein, L. B. : *Electro-Photography and Development Physics*, Springer, London, 1992.

Slatyer, R.O. : *The Physics of Structural Phase Transitions*, Academic Press, New York, 1967.

Srivastava, H.S. : *Elements of Physics*, Rastogi Publications, Meerut, 1990.

Thimann, K.V. : *Physics and the Real World*, CRC Press , Boca Raton, Florida, 1980.

Tolbert, N.E. : *Microbodies—Peroxisomes and Glyoxysomes*, Oxford University Press, London, 1971.

Toraldo, Francia G. : *The Investigation of the Physical World*, Cambridge, New York, 1981.

Verner, J.E. : *Thermal Physics*, Academic Press, New York, 1985.

Vince, Pruce D. : *Elements of Stastical and Thermodynamics Physics*, McGraw Hill, London, 1975.

Walgraif, Daniel : *Introduction to Physics*, McMillan Press, Melbourne, 1969.

Young, H. D. and Others : *University Physics*, Add. Wesley, London, 1996.

Zeevart, J.A. : *Thermal Plasma and New Materials Technology*, McGraw Hill, London, 1976.

Zhukov, L. : *Modern Physics in Twenty first Century, Springer Verlag, Berlin, 1975.*